P9-APK-024

# Developmental Biology of Peripheral Lymphoid Organs

Péter Balogh
Editor

# Developmental Biology of Peripheral Lymphoid Organs

*Editor*
Péter Balogh
University of Pécs
Department of Immunology and Biotechnology
Szigeti út 12
7624 Pécs
Hungary
peter.balogh@aok.pte.hu

ISBN 978-3-642-14428-8 e-ISBN 978-3-642-14429-5
DOI 10.1007/978-3-642-14429-5
Springer Heidelberg Dordrecht London New York



*Cover design:* WMXDesign GmbH, Heidelberg, Germany

Printed on acid-free paper

Springer is part of Springer Science+Business Media (www.springer.com)

# Preface

The human immune system is a complex network of tissues and organs dispersed throughout the body. These anatomic formations at definite locations and numbers are populated overwhelmingly with white blood cells (lymphocytes and other leukocytes) that are specialized to recognize invading pathogens and eventually destroy these.

The scene for such collaborative work is the set of tissues collectively referred to as peripheral lymphoid tissues and organs, to distinguish from those central/primary lymphoid tissues where the bulk of pathogen-responsive cells develop. Among vertebrates, the mammalians (including humans) possess the broadest range of peripheral lymphoid tissues and organs. Although similar in functions, these territories are remarkably different in the way how they emerge during development, gain functional competence, and what tissue organization they achieve.

This interlinked relationship of development–structure–functionality necessitates a volume dedicated to those developmental events that occur at the site of future immune responses, but take place prior to any encounter with external pathogens, and are crucial for subsequent immunological defense. In this regard, these biological processes strikingly mirror the evolution and advance of human society where, as a result of several thousands of years of history and social development, sophisticated infrastructure suiting highly diverse activities has been created. Buildings for living, education, work, as well as transport routes and rules have been created well before the actual need arises, but in a foreseeable and predictable pattern as a common element in preventing chaos and collapse of the system. These sites are to be filled in by people trained to perform their own individual tasks for the society's benefit as dictated by their individual capacities and conditions. In this regard, lymphocytes (demonstrating a high-degree of individuality through their clonally rearranged antigen receptors) populate and interact within those tissues whose formation had been initiated well before the duty bell rang, in the form of antigens engaging both clonal and nonclonal receptors.

This book addresses the formation of peripheral lymphoid organs without the intention of competing with excellent textbooks and other sources that are available on histology and general immunology. The objective of this book's contributors is to provide for the first time a comprehensive source in this field for those students and professionals who endeavor in studying the wonders of peripheral lymphoid organs. It is our hope that this volume will be but the first of many efforts with similar focus, and that some readers will be attracted to have a peek at the blueprint of our urban society within, so the day may come when its failures can be tackled more efficiently.

Pécs, Hungary
July 2010

Péter Balogh

# Contents

# Contributors

**Ann Ager** Department of Infection, Immunity and Biochemistry, School of Medicine, Cardiff University, Cardiff, UK

**Péter Balogh** Department of Immunology and Biotechnology, Faculty of Medicine, University of Pécs, Pécs, Hungary

**Cecile Benezech** School of Immunity and Infection, IBR-MRC Centre for Immune Regulation, College of Medical and Dental Sciences, University of Birmingham, Birmingham, UK

**Jorge H. Caamaño** School of Immunity and Infection, IBR-MRC Centre for Immune Regulation, College of Medical and Dental Sciences, University of Birmingham, Birmingham, UK

**Mark C. Coles** Centre for Immunology and Infection, Department of Biology and Hull York Medical School, University of York, York, UK

**Tom Cupedo** Department of Hematology, Erasmus University Medical Center, Rotterdam, The Netherlands

**Rania M. El Sayed** Department of Microbiology and Immunology, Virginia Commonwealth University, Richmond, VA 23298-0678, USA

**Heike Herbrand** Institute of Immunology, Hannover Medical School, Hannover, Germany

**Árpád Lábadi** Department of Immunology and Biotechnology, Faculty of Medicine, University of Pécs, Pécs, Hungary

**Peter J.L. Lane** MRC Centre for Immune Regulation, Institute for Biomedical Research, Birmingham Medical School, Birmingham, UK

**Emma Mader** School of Immunity and Infection, IBR-MRC Centre for Immune Regulation, College of Medical and Dental Sciences, University of Birmingham, Birmingham, UK

**Fiona M. McConnell** MRC Centre for Immune Regulation, Institute for Biomedical Research, Birmingham Medical School, Birmingham, UK

**Oliver Pabst** Institute of Immunology, Hannover Medical School, Hannover, Germany

**Jens V. Stein** Theodor Kocher Institute, University of Bern, Bern, Switzerland

**Andras K. Szakal** Department of Anatomy and Neurobiology, and The Immunobiology Group, Virginia Commonwealth University, Richmond, VA 23298-0678, USA

**John G. Tew** Department of Microbiology and Immunology, Virginia Commonwealth University, Richmond, VA 23298-0678, USA

**Henrique Veiga-Fernandes** Immunobiology Unit, Instituto de Medicina Molecular, Faculdade de Medicina de Lisboa, Lisboa, Portugal

**Falk Weih** Leibniz-Institute for Age Research, Fritz-Lipmann-Institute, Jena, Germany

**David Withers** MRC Centre for Immune Regulation, Institute for Biomedical Research, Birmingham Medical School, Birmingham, UK

# Chapter 1
# Introduction: Evolution of Peripheral Lymphoid Organs

**Péter Balogh**

Earlier the presence of peripheral lymphoid organs hosting immune reactions against infections was considered by lymphocyte-centered researchers as a set of tissues throughout the body being conveniently present when leukocytes need them almost akin to "deus ex machine," for either as specific homing destinations after the lymphocytes have differentiated in primary lymphohemopoietic organs or sites of immune responses. A major emphasis had been placed on the availability of lymphocytes of appropriate clonal composition and maturation status, without much consideration for the three-dimensional mesenchymal architecture of the lymphoid organs the majority of these hemopoietic cells reside in, until some key discoveries concerning the development of secondary lymphoid tissues were made. Thus following several decades of a rather quiet flow of classical embryological studies with relatively little attention in general biomedical research, however, the investigations addressing the formation of peripheral lymphoid organs have now gained a strong momentum, transforming this area into one of the most rapidly developing fields connecting developmental biology to basic and clinical immunology. Advances along three main avenues have been crucial to the renewed interest. First, our improved ability to identify minor (hemopoietic as well as stromal) cell populations by a continuously growing range of suitable markers, cell separation instruments, and procedures has greatly facilitated the characterization, high-speed purification, and subsequent analysis of cell subsets of major significance in this developmental process. Second, the expansion of procedures in genetic manipulation for targeted mutagenesis and regulated gene expression/deletion and the resulting plethora of transgenic mice have also been instrumental for revealing the role of several target cells and their progeny as well as key molecules in the process. Finally, the advances in bioinformatics with high throughput analyses have provided insight into the intracellular molecular mechanisms and developmental responses following interaction of several receptor and ligand pairs not only in physiological developmental events, but also in various pathological

P. Balogh
Department of Immunology and Biotechnology, Faculty of Medicine, University of Pécs, Pécs, Hungary

P. Balogh (ed.), *Developmental Biology of Peripheral Lymphoid Organs*,
DOI 10.1007/978-3-642-14429-5_1, 

conditions mostly associated with chronic inflammations and lymphoid malignancies. Thus, the impact of identifying key elements goes beyond understanding the physiological lymphoid organ development and its role for improved efficiency in "normal" immune responses; it may also provide opportunities to ameliorate pathological conditions related to aberrant lymphoid tissue formation.

## 1.1 Lymphoid Organ Formation and Restoration of Tissue Integrity: A Single Event or Repeating Incidents?

Depending on the mammalians species, the development of secondary lymphoid organs is completed during embryonic development (human) or shortly after birth (mouse), resulting in highly compartmentalized structure with the clustered arrangement of hemopoietic and mesenchymal cells. However, subsequent exposure to external antigens causes a substantial rearrangement of the original resting tissue architecture, affecting both sets of cells. It is now generally accepted that building efficient adaptive immune responses requires structured secondary lymphoid tissues, optimized to orchestrate the encounter of antigen-reactive lymphocytes with their cognate antigen and their subsequent expansion for transformation into effector cells. Importantly, some forms of secondary lymphoid tissues of the mucosa require a substantial exposure to bacterial influences to develop; thus, the presence of foreign antigens has a fundamental role in inducing the formation of some, but not all, secondary lymphoid tissues.

While the overwhelming majority of immunological investigations address the expansion-related changes coupled to ongoing immune responses, it seems equally important to determine how the altered architecture returns to the normal conditions during a regeneration process, once the antigen has been eliminated. Experimental evidences now indicate that this capacity of peripheral lymphoid tissues to regenerate and preserve their capacity to host immune responses in the adult period involves similar cells and largely identical molecules that promote their embryonic development. In addition to these physiological forms of tissue transformation, accumulating evidences now point to the pathological consequences of the organism's preserved capacity to establish organized lymphoid tissues in the course of chronic inflammations and organ-specific autoimmune diseases.

## 1.2 Sharing Common Purposes and Following Different Ways

Contrasting the remarkable differences between their anatomic location and antigen access, and their relative uniformity in tissue organization, it is interesting that for most peripheral lymphoid tissues that form extensive networks (lymph nodes or various lymphoid formations along the gastrointestinal tract) largely similar developmental events operate. The factors exerting the most important general effects for different lymphoid organs will be discussed in the first part of this volume. However, subtle differences are also necessary for the specification of almost each individual organ, even within the same anatomical category (for example, peripheral lymph nodes at various locations); thus, the effect of the major morphogenic compounds is modulated by a second set of regulators, often acting in an organ-specific manner. This heterogeneity in developmental requirements is best established for mammalian lymphoid organs, and ultimately it results in the broadest variety of these structures amongst vertebrates. Ultimately, these morphogenic events in each lymphoid organs result in a highly efficient three-dimensional platform for facilitating immune responses and subsequent regeneration. The specific requirements of different peripheral lymphoid organs will be presented in the second part of this volume.

# Part I
# Common Themes in Lymphoid Organ Development

The biological significance of peripheral lymphoid organ formation in higher vertebrates, including humans, is that at such sites with determined anatomical location the cells of innate and adaptive immunity may communicate and cooperate most efficiently for the elimination of external pathogens. Therefore, the presence of such highly specialized structures is crucial in establishing immunological protection against various infectious agents, contributing to the individual's (and also for the species') selection for survival. Importantly, this developmental process for most peripheral lymphoid tissues is independent not only from the exposure to external antigens (in a fashion similar to the antigen-independent formation of antigen-receptor bearing lymphocytes in primary lymphohemopoietic organs), but its initiation is also largely independent from the presence of mature T and B cells, as evidenced in mice with blocked lymphocyte development. Hence, the formation of anatomical sites committed to become secondary lymphoid organs actually predates both their reserve function for resting mature lymphocytes and also their role in focusing adaptive immune responses after birth.

Despite their macroscopic differences throughout the mammalian organisms, the tissue architecture of various peripheral lymphoid organs shows a remarkable preservation of building principles. Thus, they are strategically positioned to sample antigens from the body surfaces (regional lymph nodes), the blood circulation (spleen) and from the mucosal surface of the airways and intestinal tracts (mucosa-associated lymphoid tissues). Their tissue architectures also share notable similarities, such as separation into distinct, yet highly interchangeable, compartments with clear T or B cell dominance. The relatively stable anatomical location of these tissues paired with their similar cellular organization suggests that their formation in mammalians (where the broadest variety of such organs is generated) involves similar mechanisms, although these developmental processes manifest in a discontinuous manner, both spatially and temporally.

The first sections of this book review those cells, molecules and interaction mechanisms that are necessary for the organogenesis of secondary lymphoid tissues, and the subsequent chapters discuss the tissue-specific developmental characteristics.

# Chapter 2
# Cellular Partners in the Embryonic Induction of Lymphoid Territories: Origins and Transcriptional Regulation

**Péter Balogh**

**Abstract** A milestone in our understanding the essential steps of lymphoid organogenesis was the identification and characterization of those cell partners that are considered today as the founder pairs launching the process. This chapter summarizes the discovery and general biological features of these cells.

## 2.1 Discovery and General Characteristics of the Adam–Eve Pair of Lymphoid Organogenesis

A chance observation in neonatal mouse peripheral lymph nodes (pLNs) made in the early 1990s led to the identification of a lymphocyte subpopulation expressing the unusual phenotype of $CD4^{+}/CD3^{-}/CD45^{+}$. During the first few days immediately after birth, the frequency of these cells rapidly declines, due to the robust immigration of thymus-derived mature T cells (Kelly and Scollay 1992). These cells have no T cell generating potential and also lack cell surface immunoglobulin, although they may be induced in vitro to generate cells expressing NK cell-associated marker NK1.1 (Mebius et al. 1997).

Detailed phenotypic analysis of these cells also identified crucial markers whose activity are necessary for lymphoid organization, including IL7Rα cytokine receptor, adhesion molecules and other leukocyte markers (Mebius et al. 1997). For their seeding to developing lymph nodes $CD4^{+}/CD3^{-}/CD45^{+}$ cells utilize α4β7 integrin as counterreceptor for the MAdCAM-1 addressin (mucosal addressin cell adhesion molecule-1), transiently expressed by high endothelial venules in both peripheral and mesenteric lymph nodes (mLNs) where it will also continue to act as homing receptor in adult mice (Mebius et al. 1996). In embryos, the precursors of these cells appear as common lymphocyte progenitors (CLP) in the fetal liver and, possibly upon arrival into the fetal spleen, they acquire their $CD4^{+}/CD3^{-}$ phenotype

P. Balogh
Department of Immunology and Biotechnology, Faculty of Medicine, University of Pécs, Pécs, Hungary

P. Balogh (ed.), *Developmental Biology of Peripheral Lymphoid Organs*,
DOI 10.1007/978-3-642-14429-5_2, 

(Mebius et al. 2001). Interestingly, the original cell surface markers described for the identification of these cells do not appear to play any role in their cellular functions, as in mice lacking either CD4 or CD45 glycoprotein, the lymph nodes and spleen develop without any gross alteration of the lymphoid architecture.

Studies in murine Peyer patches (PPs) development similarly demonstrated the presence of lymphoid cells with comparable cell surface phenotype and also established that their stromal environment is enriched for mesenchymal cells expressing VCAM-1 (vascular cell adhesion molecule-1/CD106) protein. These studies led to the basic model of PP formation through sequential transformation into mature-type PP, involving the reorganization of both stromal and hemopoietic constituents within the tissue anlage in three distinct steps. In this process, each interaction between one set of hemopoietic and stromal cells creates suitable environment for subsequent colonization/survival of more differentiated hemopoietic cells, coupled with the simultaneous specification and compartmentalization of stroma (Adachi et al. 1997). These seminal observations already indicated subtle differences in the cell surface characteristics between these CD45$^+$/CD4$^+$/CD3$^-$ lymphoid cells endowed with the capacity to induce the development of both pLNs and Peyer's patches, including the largely c-Kit$^-$/CD25$^+$ phenotype for the pLN cells as opposed to the c-Kit$^+$/CD25$^-$ characteristics of PP counterpart (Nishikawa et al. 2003). Similarly to the absence of obvious need for either CD4 or CD45 molecules in lymph node development, c-kit receptor is also redundant for initiating PP organogenesis (Yoshida et al. 1999). The neonatal transfer of CD4$^+$/CD3$^-$ cells could dramatically increase the number of PPs (Finke et al. 2002), and lymphoid cells with this phenotype were subsequently referred to as lymphoid tissue inducer cells (LTi; Yoshida et al. 1999; Eberl and Littman 2003). The developmental/inducer function may probably be shared between LTi cells and related cell populations; hence, the nomenclature may also refer to some cells as lymphoid tissue initiator (LTini) cells for the latter (Ruddle and Akirav 2009). To date, the developmental features of LTini and their relationship to the LTi cells in other secondary lymphoid organs than the PP have not been thoroughly analyzed.

The studies aimed at investigating the main features of VCAM-1$^+$ mesenchymal partners of LTi cells found the expression of ICAM-1 (intercellular adhesion molecule-1) and also MAdCAM-1 adhesion molecules as cellular markers for those non-hemopoietic stromal cells that cluster with the LTi cells (Adachi et al. 1997). Analysis of highly enriched cell samples from both lymph nodes and PP demonstrated that these mesenchymal (termed lymphoid tissue organizer – LTo) cells are heterogeneous within each organ anlage with regard to the above markers (Cupedo et al. 2004a; Okuda et al. 2007), where the display of these adhesion molecules may be influenced upon interaction with LTi cells.

Once the territories committed to become lymphoid organs within the embryo have been formed, further colonization by mature lymphocytes will dictate the establishment of compartmentalized lymphoid tissue structure. This second process has two main consequences, which affects both the LTi and LTo subsets. First, the arrival of mature T and B cells will "dilute" and possibly also largely displace the LTi cells; thus, the probability and the temporal window in which lymphoid organ formation could be initiated at their normal location are probably terminated. Consequently, the mesenchymal cells are eventually exposed to a changing lymphocytic milieu, and their subsequent differentiation also reflects this alteration. Thus, B cells are required for the appearance of follicular dendritic cells (FDCs), the main non-hemopoietic compartment of the follicles (Kapasi et al. 1993), and T cells are arranged in a region enriched for T-zone fibroblastic reticular cells (FRCs), apparently also displacing the LTo population. LTi cells were demonstrated to be present in postnatal/adult murine lymphoid tissues, although at a very low frequency; but what happens to the LTo cells? More recently, a distinct FRC subpopulation situated at the peripheral rim within adult mouse lymphoid tissues was identified (Katakai et al. 2008). These cells retain the phenotypic characteristics of embryonic LTo cells, including cell surface expression of VCAM-1, ICAM-1 and MAdCAM-1 adhesion molecules, and inducible production of CXCL-13 chemokine upon engagement of lymphotoxin β-receptor (LTβR), a cardinal morphogenic member of the TNF superfamily involved in the development of all peripheral lymphoid organs (Fütterer et al. 1998). This cell population is now referred to as marginal reticular cells (MRCs), whose role in maintaining the lymphoid tissue architecture (along with the similar function of adult LTi cells) is currently under investigation. Their repositioning to the peripheral parts, such as the subcapsular sinus in pLNs and the perimeter of the white pulp in spleen (intriguingly, the main entry sites for external antigens at both locations), also indicates that the formation of lymphoid tissue architecture involves a considerable redistribution of sessile stromal cells within the developing lymphoid tissues, in addition to the continuous recirculation and migration of hemopoietic cells.

## 2.2 Transcriptional Regulation of LTi Cells

The differentiation along various hemopoietic lineages from hemopoietic stem cells simultaneously requires the ordered expression of transcription factors, and a variety of external stimuli, including soluble cytokines and membrane-bound adhesion molecules for maintaining proper cellular contacts between the developing blood cells and their stromal microenvironment. The transcription factors that guide the differentiation can either directly *promote one* type of differentiation and/or may *inhibit the divergence towards other* lineages.

Ikaros as an important member of the zinc-finger DNA-binding transcription factor family was initially demonstrated to be required for the specification of all lymphoid lineages. The deletion of N-terminal region blocks high-affinity sequence-specific interactions and causes the lack of mature T, B and NK cells in both embryonic and adult age (Georgopoulos et al. 1994). Deletion of the C-terminal domain suspends its homo- and heterodimerization necessary for high-affinity binding to DNA, and mice with mutated C-terminal domains had no recognizable pLNs and PP, in addition to the blockade of embryonic lymphoid differentiation (Wang et al. 1996). This developmental dependence of lymphoid cells that are capable of inducing the formation of pLNs and PPs on Ikaros provided further confirmation of the lymphoid affiliation of LTi cells, in addition to their lymphoid-associated cell surface marker phenotype. However, as LTi cells represent a lymphoid subset independent from both T or B cell lineage commitment (as evidenced by the presence of pLNs and PPs in mice with selective absence of either T or B cells), their differentiation pathway is controlled by a distinct set of transcription factors.

Among the crucial transcription factors that are instrumental for the differentiation of embryonic LTi cells, first the transcription factor Id2 was identified (Yokota et al. 1999). This factor (similarly to the related Id1 and Id3) binds to E proteins of basic helix-loop-helix-type (bHLH) and other transcription factors, preventing their heterodimerization. Its target genes include E2A, a key transcription factor for B cell development, and also PU.1, a dose-dependent regulator of myeloid vs. erythroid/megakaryocyte commitment. Hyperexpression of Id2 leads to an inhibition of B cell development and also facilitates its binding to PU.1 which, in turn, probably promotes the activity of GATA-1, an essential component initiating the erythroid program (Ji et al. 2008). Similarly to the lack of Ikaros, no pLNs or PPs develop in the absence of Id2, coupled with the absence of LTi cells. This lack of LTi cells is associated with the blockade of NK cell and Langerhans cell differentiation (Hacker et al. 2003).

Another selective transcription factor necessary for the development of LTi cells belongs to the retinoid-related orphan receptor (ROR) family. Members of this class of regulatory proteins are part of the nuclear hormone receptor superfamily for which the ligand has not yet been identified. These proteins contain several functional domains, engaged in (unknown) ligand binding, DNA recognition, nuclear transport and localization, and via their dimerization they can modulate gene expression, either repression or transactivation. Following ligand binding, the receptor undergoes conformational change, thus it is released from its co-repressor allowing association with co-activators. In this fashion, the interaction with the DNA target of receptor transcription factor will occur, ultimately modulating gene expression. All ROR members (RORα, RORβ and RORγ) as monomer proteins specifically recognize and bind to sequences of DNA termed ROR response elements (ROREs). Within this group, RORγt (thymus-isoform of RORγ, also expressed in double-positive thymocytes) has been identified as a crucial transcription factor for the development of LTi cells from fetal hemopoietic stem cells (Sun et al. 2000; Eberl et al. 2004).

The association of the expression of RORγt with LTi commitment has dual significance. First, it identified a transcription factor with a very restricted hemopoietic expression to be involved in the specification of these cells. Second, the expression of RORγt (studied by anti-RORγt antibodies or reporter gene such as GFP driven by RORγt promoter) could be exploited for the in situ analysis and fate mapping of LTi cells (Eberl and Littman 2004). Histological analyses demonstrated that, after their formation in the fetal liver, RORγt-positive cells in embryos first accumulate in perivascular clusters corresponding to the region of the lymphoid organ rudiments in pLNs, PPs and also the spleen, where they closely associate with

VCAM-1 positive mesenchymal cells. Similarly to the deletion of Id2, the absence of RORγt also resulted in the lack of pLNs and PPs. Importantly, the development of spleen in mice without either Id2 or RORγt is unperturbed. In addition to its probable role in survival of LTi cells, the expression of RORγt is also necessary for the formation of an invariant NKT subset in the skin and pLNs (Michel et al. 2008; Doisne et al. 2009). Interestingly, although birds also express Id2 and Ikaros and possess hemopoietic cells with NK-like activity and phenotype, yet they do not form lymph nodes (Lampisuo et al. 1999; Krishan et al. 2005).

The observations on the transcriptional requirements for LTi maturation point towards a lymphoid programming closely related to such differentiated cells that function either without clonally rearranged antigen receptors (NK cells) or antigen receptors with restricted antigenic specificity (iNKT cells), both subsets forming a numerically small compartment within the organized lymphoid tissues in adults. Therefore, it is not only the *timing* of the initiation of organ formation (predating its actual need in adaptive immune responses), but also the *identity and lineage affiliation* of LTi cells, which are difficult to reconcile with the specialized function of secondary lymphoid organs, where the antigenic specificity is a central element in adaptive immune responses, requiring highly compartmentalized tissue structure. This apparent discrepancy leaves us with an intriguing question: how and why are such uncommitted (LTi) cells produced so early, to exert their inducing function for the later benefit of other (T and B) hemopoetic cells, thus ultimately building a tissue that is not needed at the time of induction?

## 2.3 LTo Cells: From the Unknown to the Uncertain

In contrast to the emerging knowledge on the transcriptional control of LTi cells, virtually nothing is known concerning the lineage affiliation and developmental pathway of putative LTo precursors. Although cell transfer experiments have established the ectopic inducibility of organized lymphoid tissue in the early postnatal period (Cupedo et al. 2004b), it is not yet known what developmental program will dictate the transformation of uncommitted mesenchymal precursors into specified stromal elements at predisposed locations; or, if generated elsewhere, what signals cause their co-clustering with LTi cells to initiate lymphoid organ formation. Comparative studies showed that these cells have differential expression of several members of the tumor necrosis factor (TNF) superfamily and chemokines, thus providing clues for the distinct requirements for the diverse developmental pathways of pLNs, mLNs and PPs, as noted in mice with different mutations for these morphogenic regulators (Tumanov et al. 2003). Mostly due to technical challenges, much less is known about the differentiation pathway of these mesenchymal cells compared with the hemopoietic LTi cells yet. However, considering the overwhelming similarity between the LTi cells in various secondary lymphoid tissues with different developmental patterning, it is the LTo cells that have a dominant role in controlling the embryonic formation of these organs. It remains to be determined if there is any lineage relationship between the embryonic LTo cells and either the T-zone FRCs or B-zone FDCs that appear in place of the undifferentiated LTo cells as latter components (or their descendants) relocate to the peripheral segments of lymphoid territory as MRCs.

## References


Adachi S, Yoshida H, Kataoka H, Nishikawa S (1997) Three distinctive steps in Peyer's patch formation of murine embryo. Int Immunol 9:507–514

Cupedo T, Vondenhoff MF, Heeregrave EJ, De Weerd AE, Jansen W, Jackson DG, Kraal G, Mebius RE (2004a) Presumptive lymph node organizers are differentially represented in developing mesenteric and peripheral nodes. J Immunol 173:2968–2975

Cupedo T, Jansen W, Kraal G, Mebius RE (2004b) Induction of secondary and tertiary lymphoid structures in the skin. Immunity 21:655–667

Doisne JM, Becourt C, Amniai L, Duarte N, Le Luduec JB, Eberl G, Benlagha K (2009) Skin and peripheral lymph node invariant NKT cells are mainly retinoic acid receptor-related orphan receptor (gamma)t+ and respond preferentially under inflammatory conditions. J Immunol 183:2142–2149

Eberl G, Littman DR (2003) The role of the nuclear hormone receptor RORgammat in the development of lymph nodes and Peyer's patches. Immunol Rev 195:81–90

Eberl G, Littman DR. (2004) Thymic origin of intestinal alphabeta T cells revealed by fate mapping of RORgammat+ cells. Science 305:248–251

Eberl G, Marmon S, Sunshine MJ, Rennert PD, Choi Y, Littman DR (2004) An essential function for the nuclear receptor RORgamma(t) in the generation of fetal lymphoid tissue inducer cells. Nat Immunol 5:64–73

Finke D, Acha-Orbea H, Mattis A, Lipp M, Kraehenbuhl J (2002) $CD4^{+}CD3^{-}$ cells induce Peyer's patch development: role of alpha4beta1 integrin activation by CXCR5. Immunity 17:363–373

Fütterer A, Mink K, Luz A, Kosco-Vilbois MH, Pfeffer K (1998) The lymphotoxin beta receptor controls organogenesis and affinity maturation in peripheral lymphoid tissues. Immunity 9:59–70

Georgopoulos K, Bigby M, Wang JH, Molnar A, Wu P, Winandy S, Sharpe A (1994) The Ikaros gene is required for the development of all lymphoid lineages. Cell 79:143–156

Hacker C, Kirsch RD, Ju XS, Hieronymus T, Gust TC, Kuhl C, Jorgas T, Kurz SM, Rose-John S, Yokota Y, Zenke M (2003) Transcriptional profiling identifies Id2 function in dendritic cell development. Nat Immunol 4:380–386

Ji M, Li H, Suh HC, Klarmann KD, Yokota Y, Keller JR (2008) Id2 intrinsically regulates lymphoid and erythroid development via interaction with different target proteins. Blood 112:1068–1077

Kapasi ZF, Burton GF, Shultz LD, Tew JG, Szakal AK (1993) Induction of functional follicular dendritic cell development in severe combined immunodeficiency mice. Influence of B and T cells. J Immunol 150:2648–2658

Katakai T, Suto H, Sugai M, Gonda H, Togawa A, Suematsu S, Ebisuno Y, Katagiri K, Kinashi T, Shimizu A (2008) Organizer-like reticular stromal cell layer common to adult secondary lymphoid organs. J Immunol 181:6189–6200

Kelly KA, Scollay R (1992) Seeding of neonatal lymph nodes by T cells and identification of a novel population of $CD3^{-}CD4^{+}$ cells. Eur J Immunol 22:329–334

Krishan K, McKinnell I, Patel K, Dhoot GK (2005) Dynamic Id2 expression in the medial and lateral domains of avian dermamyotome. Dev Dyn 234:363–370

Lampisuo M, Liippo J, Vainio O, McNagny KM, Kulmala J, Lassila O (1999) Characterization of prethymic progenitors within the chicken embryo. Int Immunol 11:63–69

Mebius RE, Streeter PR, Michie S, Butcher EC, Weissman IL (1996) A developmental switch in lymphocyte homing receptor and endothelial vascular addressin expression regulates lymphocyte homing and permits $CD4^{+}$ $CD3^{-}$ cells to colonize lymph nodes. Proc Natl Acad Sci USA 93:11019–11024

Mebius RE, Rennert P, Weissman IL (1997) Developing lymph nodes collect $CD4^{+}CD3^{-}$ LTbeta+ cells that can differentiate to APC, NK cells, and follicular cells but not T or B cells. Immunity 7:493–504

Mebius RE, Miyamoto T, Christensen J, Domen J, Cupedo T, Weissman IL, Akashi K (2001) The fetal liver counterpart of adult common lymphoid progenitors gives rise to all lymphoid lineages, $CD45^{+}CD4^{+}CD3^{-}$ cells, as well as macrophages. J Immunol 166:6593–6601

Michel ML, Mendes-da-Cruz D, Keller AC, Lochner M, Schneider E, Dy M, Eberl G, Leite-de-Moraes MC (2008) Critical role of ROR-gammat in a new thymic pathway leading to IL-17-producing invariant NKT cell differentiation. Proc Natl Acad Sci USA 105:19845–19850

Nishikawa S, Honda K, Vieira P, Yoshida H. (2003) Organogenesis of peripheral lymphoid organs. Immunol Rev 195:72–80

Okuda M, Togawa A, Wada H, Nishikawa S. (2007) Distinct activities of stromal cells involved in the organogenesis of lymph nodes and Peyer's patches. J Immunol 179:804–811

Ruddle NH, Akirav EM (2009) Secondary lymphoid organs: responding to genetic and environmental cues in ontogeny and the immune response. J Immunol 183:2205–2212

Sun Z, Unutmaz D, Zou YR, Sunshine MJ, Pierani A, Brenner-Morton S, Mebius RE, Littman DR (2000) Requirement for RORgamma in thymocyte survival and lymphoid organ development. Science 288:2369–2373

Tumanov AV, Grivennikov SI, Shakhov AN, Rybtsov SA, Koroleva EP, Takeda J, Nedospasov SA, Kuprash DV (2003) Dissecting the role of lymphotoxin in lymphoid organs by conditional targeting. Immunol Rev 195:106–116

Wang JH, Nichogiannopoulou A, Wu L, Sun L, Sharpe AH, Bigby M, Georgopoulos K (1996) Selective defects in the development of the fetal and adult lymphoid system in mice with an Ikaros null mutation. Immunity 5:537–549

Yokota Y, Mansouri A, Mori S, Sugawara S, Adachi S, Nishikawa S, Gruss P (1999) Development of peripheral lymphoid organs and natural killer cells depends on the helix-loop-helix inhibitor Id2. Nature 397:702–706

Yoshida H, Honda K, Shinkura R, Adachi S, Nishikawa S, Maki K, Ikuta K, Nishikawa SI (1999) IL-7 receptor $alpha^{+}$ CD3(−) cells in the embryonic intestine induces the organizing center of Peyer's patches. Int Immunol 11:643–655

# Chapter 3
# Lymphotoxin/Tumour Necrosis Factor Family Members as Morphogenic Factors

**Péter Balogh**

**Abstract** Initially described as potent soluble mediators involved in inflammation, Tumour Necrosis Factor (TNF) and lymphotoxin alpha and beta (LTα/β) are now also appreciated as cardinal morphogenic regulators in secondary lymphoid organ formation, in addition to related activities of other members of the same TNF/LT family. Their developmental effects simultaneously show both substantial overlaps and remarkable tissue-specific differences. This part details the receptor–ligand relationship and developmental roles of those members of this large group of soluble or membrane-bound cytokines that have been associated with secondary lymphoid organ formation.

## 3.1 Introduction: TNF and LT – from Inflammation and Cancer to Central Stage of Lymphoid Organogenesis

One of the most remarkable advances in our understanding the process of lymphoid organogenesis has been the gradual shift in appreciating the role of a group of proteins whose first identified members as soluble mediators, tumour necrosis factor (TNF), and the closely related lymphotoxin (LT) were initially found to induce inflammation and hemorrhagic necrosis of certain tumours and were cloned (Pennica et al. 1984; Gray et al. 1984; Beutler et al. 1985; Beutler and Cerami 1989). In addition, this continuing transformation of understanding the features of these molecules and their receptors has also led to the development and therapeutic use of some of the most effective anti-inflammatory agents, thus also revolutionizing the concept of immune suppression in chronic inflammations and autoimmune disorders.

These first members of a family now encompassing over 40 members, including ligands and receptors (Bodmer et al. 2002; Hehlgans and Pfeffer 2005), were initially referred to as TNFα and LT or TNFβ. Their genes are located within the

P. Balogh
Department of Immunology and Biotechnology, Faculty of Medicine, University of Pécs, Pécs, Hungary

P. Balogh (ed.), *Developmental Biology of Peripheral Lymphoid Organs*,
DOI 10.1007/978-3-642-14429-5_3, 

gene complex encoding MHC in both mouse and man on chromosomes 17 and 6, respectively (Müller et al. 1987). Importantly, in addition to their soluble forms, both TNF and LT were later found to be expressed on the surface of activated lymphoid cells (Ware et al. 1992). TNF is recognized by two types of TNF receptors, TNF-R1/p55 (CD120a) and TNF-R2/p75 (CD120b) (Brockhaus et al. 1990) that also bind LT (Hohmann et al. 1990). Subsequently, another component associated with LT/TNFβ was found to be expressed as part of a cell surface heterotrimeric complex (Browning et al. 1993), thus necessitating another renaming: TNFα remained TNF, while LT/TNFβ was denoted as LTα, and the associating partner became known as LTβ. Gene expression studies demonstrated that LTβ is constitutively expressed in the white pulp of spleen (Pokholok et al. 1995). In heterotrimeric complex, the dominant cell-bound form is composed of LTα/LTβ2 (Browning et al. 1995), where one (soluble) LTα is non-covalently paired with two (transmembrane) LTβ subunits, recognized by its receptor LTβR (Crowe et al. 1994), although a minor fraction composed of LTα2/LTβ may also bind to TNF-R1 (Ware et al. 1995). In the dominant LTα/β2 trimeric form, the LTα component is necessary for the complex formation, but the specific receptor binding to LTβR is mediated by the two LTβ constituents (Williams-Abbott et al. 1997).

The first indication that TNF/LT molecules have important roles in the formation of peripheral lymphoid organs was the profound alteration of normal lymphoid structure of spleen and the absence of lymph nodes in mice lacking LTα (De Togni et al. 1994). Subsequent studies using a broad range of transgenic and knock-out mice investigated the various roles these TNF/LT family members play in orchestrating the development of secondary lymphoid organs. The use of mice rendered deficient for either the ligand(s) or receptor is also often combined with reciprocal bone marrow transplantation, in order to dissect the impact of hemopoietic and non-transplantable elements in inducing/correcting lymphoid architecture as dictated by their expression pattern of ligands/receptors.

## 3.2 Non-overlapping Effects of TNF and LT Recognition in Secondary Lymphoid Organ Development

Earlier biochemical analyses already indicated a substantial overlap between various LT/TNF family members for receptor binding (Bodmer et al. 2002; Hehlgans and Pfeffer 2005); thus, predictably, the use of KO mice for studying the effects of select deficiencies of ligand/receptor family members could only have resulted in conditions with equally complex interpretation. For example, a mouse mutant also rendered deficient for LTα demonstrated the presence of rudimentary mLNs (Banks et al. 1995), despite the earlier report for the absence of these lymphoid tissues (De Togni et al. 1994). Since LTα may function as soluble homotrimer or part of a cell-bound heterotrimer with LTβ, it was important to determine in what actual form LTα acts during organogenesis. In mice rendered deficient for LTβ using either

Cre-Lox-mediated excision (Alimzhanov et al. 1997) or with deleted LTβ gene (Koni et al. 1997) where only soluble LTα3 homotrimer was available for binding to TNF-R1 and TNF-R2, largely similar alterations to those observed in LTα deficiency were found. However, cervical and mLNs were formed, indicating that at these locations lymphoid organogenesis may proceed independently from the presence of LTα-LTβ complexes (thus also including LTα2/β and LTα/β2 as complex ligands). However, in mice deficient for LTβR, no lymph node development (including cervical and mesenteric) could be detected, in addition to the severely perturbed white pulp architecture. This finding raised the possibility of alternative ligand(s) for LTβR other than LTα/β2 (Fütterer et al. 1998) that can engage LTβR even in either LTα- or LTβ-deficient mice. As a possible candidate, another soluble molecule, LIGHT (*l*ymphotoxins, exhibits *i*nducible expression, and competes with HSV *g*lycoprotein D for *H*VEM, a receptor expressed by T lymphocytes), a ligand for herpes virus entry mediator (HVEM) and for LTβR (Mauri et al. 1998) was found to be involved in supplementing some stimuli for LTβR, as evidenced in the lower frequency of mice developing mLN in LTβ/LIGHT double-mutant mice compared with LTβ KO only, although even the combined absence of LIGHT and LTβ could not completely block the formation of mLNs. Furthermore, the resulting spleen architecture in these double-KO mice was less severely perturbed than in LTβR-deficient mice, raising the possibility of yet another unidentified ligand for LTβR (Scheu et al. 2002). Now TNF, TNF-R1/2, LTα, LTβ, LTβR, LIGHT and HVEM of the TNF ligand/receptor superfamily are referred to as the immediate TNF family members (Ware 2008).

As in mice rendered deficient for LTβR, the defects manifest since the earliest time point for expressing this molecule, another approach was necessary to determine the temporal regulation of the initiation of lymphoid organogenesis. The in vivo administration of a soluble decoy receptor composed of the LTβR fused to human IgG (LTβR-huIgG) into timed pregnant mice revealed that, depending on the time of injection, the formation of various pLNs could be blocked, albeit the development of mLNs could not be influenced even with treatment starting as early as day 9 of pregnancy. Thus, the presence of lymph nodes in mice born after even early in utero treatment suggests that the induction of some lymph node formation is independent from LTα/β. Of the pLNs, the brachial and axillary LNs could be blocked by injection of fusion protein up to gestational age day 12 and 14, respectively, whereas popliteal LN formation could be inhibited with LTβR-huIgG injection up to day 16, thus outlining a cranio-caudal order of LN development (Rennert et al. 1996). In a reverse experiment, in mice with LTα-deficiency (with no peripheral lymph nodes), the administration of agonist anti-LTβR monoclonal antibody could efficiently induce the formation of LNs, again, with their anatomical location corresponding to a temporal order of treatment. However, the architecture of these antibody-induced LN-like aggregates did not reach adult-type compartmentalization, presumably due to the absence of LTα effects exerted upon TNF-R1 and TNF-R2. Correspondingly, complete blockade of TNF-R1 and TNF-R2 in conjunction with the inhibition of LTβR signalling resulted in the inhibition of all lymph nodes, including those that were resistant to the absence of LTα or treatment with LTβR-huIgG decoy receptor (Rennert et al. 1998). It is not yet known how the lack of LTβR alone in KO mice (Fütterer et al. 1998) is sufficient to achieve as extensive a blockade of lymph node genesis as the intervention combining the antibody-mediated blockade of known receptors for both TNF/LTα and LTβ.

In addition to the variable effects of the absence of LTα or LTβ on the formation of lymph nodes at various anatomical regions, Peyer's patches are uniformly blocked in the absence of these ligands, while they are present with disrupted LIGHT (Pfeffer 2003). However, in contrast to the inducibility of (disorganized) lymph nodes in LTα-deficient mice treated in utero with agonist anti-LTβR antibody, the development of Peyer's patches could not be promoted (Rennert et al. 1998). The combined inactivation of the entire set of TNF/LTα/LTβ resulted in a composite blockade in the formation of peripheral lymphoid organs, which was more severe than either the isolated LTβ or TNF/LTβ double deficient mutant, indicating non-redundant roles for these cytokines (Kuprash et al. 2002). Importantly, selective deficiency of LTβ from various lymphocyte subsets also established that, depending on the type of peripheral lymphoid tissue, distinct lymphocyte subpopulation-associated expression of LTβ was required for tissue formation. Thus in pLNs and, to a lesser extent, Peyer's patches, the lack of LTβ in B cells or T cells did not influence the tissue architecture, whereas the selective lack of LTβ in B cells severely altered the organization of spleen, although not as extensively as in complete LTβ-deficiency; therefore, T cell-associated LTβ also may provide auxiliary role(s) in this organ (Tumanov et al. 2003). Although the lack of either LTα or LTβR permits the formation of the spleen as separate organ, it results in the absence of marginal zone and severely perturbed white pulp architecture, including the absence of FDCs and T-B compartmentalization (De Togni et al. 1994; Alimzhanov et al. 1997; Fütterer et al. 1998). Nevertheless, the segregation of red pulp and white pulp as two major domains of the spleen is preserved.

Compared with the profound inhibition of lymph node and Peyer's patches formation in the absence of LTαβ and LTβR, the morphogenic consequences of the lack of TNF or TNF receptors TNF-R1 and TNF-R2 are much more subtle. In mice with disrupted TNF-R2, no deficiency or structural alteration of peripheral lymphoid organs were observed. Lymph nodes and Peyer's patches are also formed in the absence of TNF; however, no germinal centres can be induced and FDCs are also absent (Pasparakis et al. 1996). In the absence of TNF-R1, the development of Peyer's patches is also blocked, although lymph nodes form (Pasparakis et al. 1997). Thus it appears that, in a sequential order following the LTα and LTβ-directed establishment of lymphoid territories, TNF and its receptor (TNF-R1) act for further cellular differentiation and tissue maturation, consequently enabling the peripheral lymphoid organs to mount efficient immune responses.

Further to the effects of the absence of TNF on lymphoid organogenesis, the deficiency of Type I interferon family member IFNβ also affects splenic architecture to a moderate degree. Possibly via the induction of TNFα, IFNβ indirectly influences T/B zone segregation and red pulp macrophage colonization. The disorganized lymphoid structure of spleen in IFNβ KO mice suggests that, in this function, other members of this family (IFNα, IFNω and IFNκ) cannot replace IFNβ, despite binding to the same receptors (Deonarain et al. 2003).

## 3.3 Role of TRANCE in Lymphoid Organogenesis

In addition to LTs, another ligand–receptor pair of the LT/TNF family TRANCE (RANKL) and its receptor TRANCE-R/RANK were also found necessary for the formation of lymph nodes, without influencing the generation of Peyer's patches and spleen, even though some structural defects with variable severity were observed in spleen (Kong et al. 1999; Dougall et al. 1999; Kim et al. 2000). TRANCE in lymph nodes from newborn mice was found to be expressed by cells displaying $CD45^{+}/CD3^{-}/CD4^{+}$ LTi phenotype, together with LTαβ2 and also TRANCE-R (RANK), whereas LTβR was expressed only by the non-hemopoietic compartment. In developing lymph nodes, the TRANCE-RANK mechanism appears to regulate the fate of organ anlage by priming LTi cells to upregulate LTαβ2 expression, thus creating a positive feedback for augmented LTβR stimulation in immature stromal organizer cells, leading to their increased production of soluble factors, such as IL-7 and TRANCE (Vondenhoff et al. 2009). Interestingly, the in vivo administration of IL-7 to embryonic mice with disrupted RANK-induced signalling could restore the clustering of stromal precursors and initiation of lymph node formation, although the structure of these tissues was abnormal (Yoshida et al. 2002). This compartmentalization of different bioavailable stimulants (IL-7 available within Peyer's patches anlage) also points to the impact of tissue milieu on the survival and developmental potential of LTi cells, where the effect of soluble stimulants may, at least partially, substitute for the local absence of membrane-bound TRANCE ligands.

## 3.4 Signalling Mechanisms of the Immediate TNF Family Members: A Net of TRAFs and Others

The typical mechanism of ligand recognition within the TNF family is the association of trimeric type II transmembrane ligands (cell-bound or soluble) with type I transmembrane receptors, also typically trimeric protein complexes. Although the pattern of partially overlapping ligand binding by LT/TNF receptors suggests some redundancy, the consequence of their absence demonstrates non-redundant functions, including cell proliferation, survival, differentiation and apoptosis of responding cells, with profound impact in lymphoid organ development as well as subsequent immune responses. In addition, several interacting partners of ligands and receptors utilize similar signalling mechanisms; thus, a substantial integration of signals is apparent at both receptor-ligand level and downstream intracellular events. The two main outcomes at cellular level is either the induction of apoptosis or cell activation, latter involving two alternate pathways of NF-κB processing, JNK, ERK, AP-1 and other transcription factor-mediated gene expression pathways (Hehlgans and Pfeffer 2005) and, as third effect, some TNF family receptors also perform decoy functions. Due to its crucial significance, various aspects of NF-κB signalling pathways and their roles in lymphoid organogenesis are presented in Chap. 4.

The majority of immediate LT/TNF receptors (TNF-R2, LTβR and HVEM) and also TRANCE-R/RANK contain conserved TIM motifs interacting with TNF-associated factors (TRAF-1,2,3,5 and TRAF-6), whereas the cytoplasmic part of TNF-R1 contains death domain (Ha et al. 2009). The physiological role of TRAF-4 in TNF family signalling is currently unknown (Cherfils-Vicini et al. 2008).

The involvement of various TRAFs in distinct LT/TNF receptor functions revealed a highly complex system in signalling. Similarly to the ligand and receptor trimeric structure, TRAFs also oligomerize both in homotrimeric or heterotrimeric forms. The TRAF protein oligomerization occurs through the TRAF domains that locate at the C-terminal part of the proteins, also interacting with the TNF receptors, whereas the Zn-finger elements at the N-terminal region are important for activating kinases upon association of TRAF proteins with ligand-bound receptor. TNF-R1 may indirectly interact with TRAF-1,2 and TRAF-5, whereas TRAF-6 may also couple with IL-1 receptor via the involvement of IL-1R-associated kinase IRAK (Cao et al. 1996), and may transmit signals from other receptors, including IL-18 receptor and Toll-like receptors (TLRs). Although the association of TRAFs with various LT/TNF receptors has been established, the precise pattern and signalling involvement has not been completely explored yet.

While TRAF2 and TRAF5 may transmit LTβR-derived signals, their absence does not lead to the blockade of lymph node formation as observed in the absence of LTβR deficiency. On the other hand, the formation of Peyer's patches is severely blocked in TRAF2-deficient mice, together with impaired signalling of TNFR1, indicating that TRAF2 may transmit signals that are necessary for the Peyer's patches formation primarily via associating with TRAF2 (Piao et al. 2007). In contrast, the signal elicited by TRANCE binding by TRANCE-R/RANK is likely to be mediated through TRAF6, as mice with ablated TRAF6 also have a severe defect of lymph node development (Naito et al. 1999). In the spleen of TRAF6-deficient mice, the initial formation of follicles together with the differentiation of FDCs occurred, although its completion was blocked in the late neonatal period, similarly to a blockade of follicle development in Peyer's patches (Qin et al. 2007). The systemic lack of TRAF3 causes early postnatal lethality, and its absence did not lead to any detectable tissue alteration of the spleen (Xu et al. 1996). Its targeted inactivation selectively in B cells, however, leads to an increased B cell proliferation and antibody production, illustrating its role in limiting B cell expansion and activation under physiological conditions (Xie et al. 2007).

As many ligands of the LT/TNF family are also transmembrane proteins, it raises the possibility of "reverse signalling", i.e. when the ligands themselves also serve as sensors. In this fashion TRANCE, and TNF as important members of the

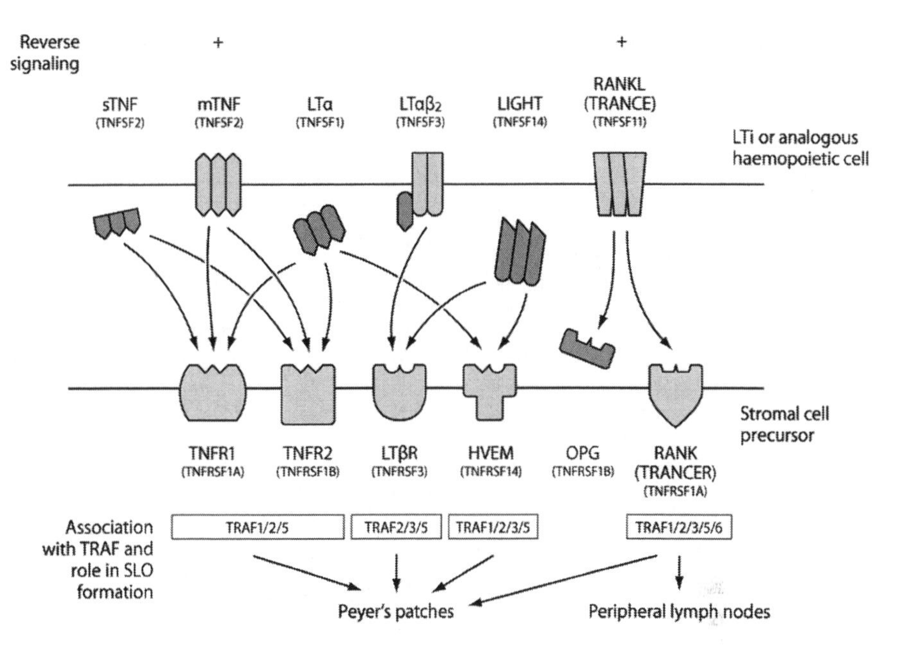


**Fig. 3.1** The chart depicts those ligand–receptor interactions of LT/TNF family members and their downstream TRAF-signalling elements that promote secondary lymphoid organ (SLO) formation acting on various receptors expressed by stromal cells or their precursors (abbreviations are listed in the text; in parenthesis the corresponding TNFSF/TNFRSF nomenclature is used). Some transmembrane ligand members of the family may exert reverse signalling towards the hemopoietic cells (LTi or analogous subset)

immediate LT/TNF family involved in lymphoid organogenesis have been proposed to exert such activities in lymphoid cells, whereas other members may influence immune responses via this mechanism by modulating other stimulatory effects (Eissner et al. 2004; Sun and Fink 2007).

Figure 3.1 summarizes the ligand–receptor relationship of immediate LT/TNF members, their TRAF-association and potential reverse signalling activities.

## References

Alimzhanov MB, Kuprash DV, Kosco-Vilbois MH, Luz A, Turetskaya RL, Tarakhovsky A, Rajewsky K, Nedospasov SA, Pfeffer K. (1997) Abnormal development of secondary lymphoid tissues in lymphotoxin beta-deficient mice. Proc Natl Acad Sci USA 94:9302–9307

Banks TA, Rouse BT, Kerley MK, Blair PJ, Godfrey VL, Kuklin NA, Bouley DM, Thomas J, Kanangat S, Mucenski ML (1995) Lymphotoxin-alpha-deficient mice. Effects on secondary lymphoid organ development and humoral immune responsiveness. J Immunol 155:1685–1693

Beutler B, Cerami A (1989) The biology of cachectin/TNF – a primary mediator of the host response. Annu Rev Immunol 7:625–655

Beutler B, Greenwald D, Hulmes JD, Chang M, Pan YC, Mathison J, Ulevitch R, Cerami A (1985) Identity of tumour necrosis factor and the macrophage-secreted factor cachectin. Nature 316:552–554

Bodmer JL, Schneider P, Tschopp J (2002) The molecular architecture of the TNF superfamily. Trends Biochem Sci 27:19–26

Brockhaus M, Schoenfeld HJ, Schlaeger EJ, Hunziker W, Lesslauer W, Loetscher H. (1990) Identification of two types of tumor necrosis factor receptors on human cell lines by monoclonal antibodies. Proc Natl Acad Sci USA 87:3127–3131

Browning JL, Ngam-ek A, Lawton P, DeMarinis J, Tizard R, Chow EP, Hession C, O'Brine-Greco B, Foley SF, Ware CF (1993) Lymphotoxin beta, a novel member of the TNF family that forms a heteromeric complex with lymphotoxin on the cell surface. Cell 72:847–56

Browning JL, Dougas I, Ngam-ek A, Bourdon PR, Ehrenfels BN, Miatkowski K, Zafari M, Yampaglia AM, Lawton P, Meier W et al (1995) Characterization of surface lymphotoxin forms. Use of specific monoclonal antibodies and soluble receptors. J Immunol 154:33–46

Cao Z, Xiong J, Takeuchi M, Kurama T, Goeddel DV (1996) TRAF6 is a signal transducer for interleukin-1. Nature 383:443–446

Cherfils-Vicini J, Vingert B, Varin A, Tartour E, Fridman WH, Sautès-Fridman C, Régnier CH, Cremer I. (2008) Characterization of immune functions in TRAF4-deficient mice. Immunology 124:562–574

Crowe PD, VanArsdale TL, Walter BN, Ware CF, Hession C, Ehrenfels B, Browning JL, Din WS, Goodwin RG, Smith CA (1994) A lymphotoxin-beta-specific receptor. Science 264:707–710

De Togni P, Goellner T, Ruddle NH, Streeter PR, Fick A, Mariathasan S, Smith SC, Carlson R, Shornick LP, Strauss-Schoenberger J, Russell JH, Karr R, Chaplin DD (1994) Abnormal development of peripheral lymphoid organs in mice deficient in lymphotoxin. Science 264:703–707

Deonarain R, Verma A, Porter AC, Gewert DR, Platanias LC, Fish EN. (2003) Critical roles for IFN-beta in lymphoid development, myelopoiesis, and tumor development: links to tumor necrosis factor alpha. Proc Natl Acad Sci USA 100:13453–13458

Dougall WC, Glaccum M, Charrier K, Rohrbach K, Brasel K, De Smedt T, Daro E, Smith J, Tometsko ME, Maliszewski CR, Armstrong A, Shen V, Bain S, Cosman D, Anderson D, Morrissey PJ, Peschon JJ, Schuh J (1999) RANK is essential for osteoclast and lymph node development. Genes Dev 13:2412–2424

Eissner G, Kolch W, Scheurich P (2004) Ligands working as receptors: reverse signaling by members of the TNF superfamily enhance the plasticity of the immune system. Cytokine Growth Factor Rev 15:353–366

Fütterer A, Mink K, Luz A, Kosco-Vilbois MH, Pfeffer K. (1998) The lymphotoxin beta receptor controls organogenesis and affinity maturation in peripheral lymphoid tissues. Immunity 9:59–70

Gray PW, Aggarwal BB, Benton CV, Bringman TS, Henzel WJ, Jarrett JA, Leung DW, Moffat B, Ng P, Svedersky LP et al (1984) Cloning and expression of cDNA for human lymphotoxin, a lymphokine with tumour necrosis activity. Nature 312:721–724

Ha H, Han D, Choi Y (2009) TRAF-mediated TNFR-family signaling. Curr Protoc Immunol Nov; Chapter 11:Unit11.9D

Hehlgans T, Pfeffer K (2005) The intriguing biology of the tumour necrosis factor/tumour necrosis factor receptor superfamily: players, rules and the games. Immunology 115:1–20

Hohmann HP, Remy R, Pöschl B, van Loon AP (1990) Tumor necrosis factors-alpha and -beta bind to the same two types of tumor necrosis factor receptors and maximally activate the transcription factor NF-kappa B at low receptor occupancy and within minutes after receptor binding. J Biol Chem 265:15183–15188

Kim D, Mebius RE, MacMicking JD, Jung S, Cupedo T, Castellanos Y, Rho J, Wong BR, Josien R, Kim N, Rennert PD, Choi Y (2000) Regulation of peripheral lymph node genesis by the tumor necrosis factor family member TRANCE. J Exp Med 192:1467–1478

Kong YY, Yoshida H, Sarosi I, Tan HL, Timms E, Capparelli C, Morony S, Oliveira-dos-Santos AJ, Van G, Itie A, Khoo W, Wakeham A, Dunstan CR, Lacey DL, Mak TW, Boyle WJ, Penninger JM. (1999) OPGL is a key regulator of osteoclastogenesis, lymphocyte development and lymph-node organogenesis. Nature 397:315–323

Koni PA, Sacca R, Lawton P, Browning JL, Ruddle NH, Flavell RA. (1997) Distinct roles in lymphoid organogenesis for lymphotoxins alpha and beta revealed in lymphotoxin beta-deficient mice. Immunity 6:491–500

Kuprash DV, Alimzhanov MB, Tumanov AV, Grivennikov SI, Shakhov AN, Drutskaya LN, Marino MW, Turetskaya RL, Anderson AO, Rajewsky K, Pfeffer K, Nedospasov SA (2002) Redundancy in tumor necrosis factor (TNF) and lymphotoxin (LT) signaling in vivo: mice with inactivation of the entire TNF/LT locus versus single-knockout mice. Mol Cell Biol 22:8626–8634

Mauri DN, Ebner R, Montgomery RI, Kochel KD, Cheung TC, Yu GL, Ruben S, Murphy M, Eisenberg RJ, Cohen GH, Spear PG, Ware CF (1998) LIGHT, a new member of the TNF superfamily, and lymphotoxin alpha are ligands for herpesvirus entry mediator. Immunity 8:21–30

Müller U, Jongeneel CV, Nedospasov SA, Lindahl KF, Steinmetz M (1987) Tumour necrosis factor and lymphotoxin genes map close to H-2D in the mouse major histocompatibility complex. Nature 325:265–267

Naito A, Azuma S, Tanaka S, Miyazaki T, Takaki S, Takatsu K, Nakao K, Nakamura K, Katsuki M, Yamamoto T, Inoue J (1999) Severe osteopetrosis, defective interleukin-1 signalling and lymph node organogenesis in TRAF6-deficient mice. Genes Cells 4:353–362

Pasparakis M, Alexopoulou L, Episkopou V, Kollias G (1996) Immune and inflammatory responses in TNF alpha-deficient mice: a critical requirement for TNF alpha in the formation of primary B cell follicles, follicular dendritic cell networks and germinal centers, and in the maturation of the humoral immune response. J Exp Med 184:1397–1411

Pasparakis M, Alexopoulou L, Grell M, Pfizenmaier K, Bluethmann H, Kollias G (1997) Peyer's patch organogenesis is intact yet formation of B lymphocyte follicles is defective in peripheral lymphoid organs of mice deficient for tumor necrosis factor and its 55-kDa receptor. Proc Natl Acad Sci USA 94:6319–6323

Pennica D, Nedwin GE, Hayflick JS, Seeburg PH, Derynck R, Palladino MA, Kohr WJ, Aggarwal BB, Goeddel DV (1984) Human tumour necrosis factor: precursor structure, expression and homology to lymphotoxin. Nature 312:724–729

Pfeffer K (2003) Biological functions of tumor necrosis factor cytokines and their receptors. Cytokine Growth Factor Rev 14:185–191

Piao JH, Yoshida H, Yeh WC, Doi T, Xue X, Yagita H, Okumura K, Nakano H (2007) TNF receptor-associated factor 2-dependent canonical pathway is crucial for the development of Peyer's patches. J Immunol 178:2272–2277

Pokholok DK, Maroulakou IG, Kuprash DV, Alimzhanov MB, Kozlov SV, Novobrantseva TI, Turetskaya RL, Green JE, Nedospasov SA (1995) Cloning and expression analysis of the murine lymphotoxin beta gene. Proc Natl Acad Sci USA 92:674–678

Qin J, Konno H, Ohshima D, Yanai H, Motegi H, Shimo Y, Hirota F, Matsumoto M, Takaki S, Inoue J, Akiyama T (2007) Developmental stage-dependent collaboration between the TNF receptor-associated factor 6 and lymphotoxin pathways for B cell follicle organization in secondary lymphoid organs. J Immunol 179:6799–6807

Rennert PD, Browning JL, Mebius R, Mackay F, Hochman PS. (1996) Surface lymphotoxin alpha/beta complex is required for the development of peripheral lymphoid organs. J Exp Med 184:1999–2006

Rennert PD, James D, Mackay F, Browning JL, Hochman PS (1998) Lymph node genesis is induced by signaling through the lymphotoxin beta receptor. Immunity 9:71–79

Scheu S, Alferink J, Pötzel T, Barchet W, Kalinke U, Pfeffer K (2002) Targeted disruption of LIGHT causes defects in costimulatory T cell activation and reveals cooperation with lymphotoxin beta in mesenteric lymph node genesis. J Exp Med 195:1613–1624

Sun M, Fink PJ (2007) A new class of reverse signaling costimulators belongs to the TNF family. J Immunol 179:4307–4312

Tumanov AV, Grivennikov SI, Shakhov AN, Rybtsov SA, Koroleva EP, Takeda J, Nedospasov SA, Kuprash DV (2003) Dissecting the role of lymphotoxin in lymphoid organs by conditional targeting. Immunol Rev 195:106–116

Vondenhoff MF, Greuter M, Goverse G, Elewaut D, Dewint P, Ware CF, Hoorweg K, Kraal G, Mebius RE (2009) LTbetaR signaling induces cytokine expression and up-regulates lymphangiogenic factors in lymph node anlagen. J Immunol 182:5439–5445

Ware CF (2008) Targeting lymphocyte activation through the lymphotoxin and LIGHT pathways. Immunol Rev 223:186–201

Ware CF, Crowe PD, Grayson MH, Androlewicz MJ, Browning JL (1992) Expression of surface lymphotoxin and tumor necrosis factor on activated T, B, and natural killer cells. J Immunol 149:3881–3888

Ware CF, VanArsdale TL, Crowe PD, Browning JL (1995) The ligands and receptors of the lymphotoxin system. Curr Top Microbiol Immunol 198:175–218

Williams-Abbott L, Walter BN, Cheung TC, Goh CR, Porter AG, Ware CF (1997) The Lymphotoxin-alpha (LTalpha) subunit is essential for the assembly, but not for the receptor specificity, of the membrane-anchored LTalpha 1beta 2 heterotrimeric ligand. J Biol Chem 272:19451–19456

Xie P, Stunz LL, Larison KD, Yang B, Bishop GA (2007) Tumor necrosis factor receptor-associated factor 3 is a critical regulator of B cell homeostasis in secondary lymphoid organs. Immunity 27:253–267

Xu Y, Cheng G, Baltimore D (1996) Targeted disruption of TRAF3 leads to postnatal lethality and defective T-dependent immune responses. Immunity 5:407–415

Yoshida H, Naito A, Inoue J, Satoh M, Santee-Cooper SM, Ware CF, Togawa A, Nishikawa S, Nishikawa S (2002) Different cytokines induce surface lymphotoxin-alphabeta on IL-7 receptor-alpha cells that differentially engender lymph nodes and Peyer's patches. Immunity 17:823–833

# Chapter 4
# NF-κB Signalling and Lymphoid Tissue Organogenesis

**Cecile Benezech, Emma Mader, Falk Weih, and Jorge Caamaño**

**Abstract** The development of secondary lymphoid organs is initiated by crosstalk interactions between bone marrow-derived cells and stromal cells. Several of the molecules mediating this cell–cell communication are members of the Tumour Necrosis Factor (TNF) family of ligands and receptors that induce changes in the gene expression program through activation of different members of the Nuclear Factor kappa B (NF-κB) family of transcription factors. Engagement of TNF receptors by their specific ligands results in the activation of the canonical and non-canonical NF-κB pathways that ultimately induce the expression of cytokines, chemokines and cell adhesion molecules required for lymphorganogenesis, organization of specific areas in these organs and their long-term maintenance. Analysis of transgenic mice with impairments in these pathways have underlined the important function of different NF-κB family members, their cell specificity and requirement during lymphoid tissue organogenesis and helped to define some of their transcriptional targets.

## 4.1 Introduction

Secondary lymphoid organs (SLO) such as lymph nodes (LN) contain highly organized structures including cortical B cell follicles, containing re-circulating naïve B cells searching for their specific antigen, and para-cortical T cell areas containing dendritic cells (DC) and T lymphocytes. The organization of para-cortical areas facilitates the interaction between DC presenting antigen and T cells searching for their cognate peptide through the T cell receptor and thus

C. Benezech, E. Mader, and J. Caamaño (✉)
School of Immunity and Infection, IBR-MRC Centre for Immune Regulation, College of Medical and Dental Sciences, University of Birmingham, Birmingham, UK
e-mail: J.Caamano@bham.ac.uk

F. Weih
Leibniz-Institute for Age Research, Fritz-Lipmann-Institute, Jena, Germany

P. Balogh (ed.), *Developmental Biology of Peripheral Lymphoid Organs*,
DOI 10.1007/978-3-642-14429-5_4, 

initiating adaptive immune responses. At the core of the organization of the B and T cell areas are different types of stromal cells, expressing B or T cell attracting chemokines, and forming a network of conduits that allows lymphocytes to move to the appropriate microenvironment which facilitate cell–cell interactions (Mueller and Germain 2009).

Members of the Tumour Necrosis Factor (TNF) family of ligands and receptors such as Lymphotoxin Alfa (LTα) and Lymphotoxin Beta (LTβ), and their Lymphotoxin Beta Receptor (LTβR), Receptor Activator of NF-κB ligand (RANK-L/TRANCE) and its receptor RANK, and TNFα and TNF-R1 have been shown to be required for the development of SLO during embryogenesis and for their maintenance and the organization of specific lymphocyte areas (Fu and Chaplin 1999; Mebius 2003; Randall et al. 2008). Signalling downstream of the TNF-R family members mentioned above induces the activation and nuclear translocation of the Nuclear Factor kappa B (NF-κB) family of transcription factors. The aim of this chapter is to integrate the current understanding of the role of the different NF-κB family members in the development and organization of secondary lymphoid tissues.

## 4.2 Activation of the NF-κB Transcription Factors

The NF-κB proteins were described for the first time more than 20 years ago as possessing a DNA binding activity that induced the expression of the immunoglobulin kappa light chain. The work of a large number of groups has shown that these transcription factors are essential in innate and adaptive immune responses, mediating cell survival as well as cell death, organogenesis, maturation and differentiation of different cell lineages and cell transformation (Perkins 2007; Ghosh and Hayden 2008; Vallabhapurapu and Karin 2009).

The NF-κB family of proteins in vertebrates contains five members: NF-κB1, NF-κB2, RelA, RelB and c-Rel (Fig. 4.1). Common to all NF-κB family members is a region called Rel homology domain that contains the DNA binding region, the dimerization domain and the nuclear localization signal. NF-κB1 (p105) and NF-κB2 (p100) are synthesized as large precursors that upon cell stimulation and limited processing by the proteasome generate the DNA binding subunits NF-κB1 p50 and NF-κB2 p52. Lack of transcriptional activation domains in p50 and p52 means that their homodimers will bind DNA and repress gene expression. However, when they form heterodimers with RelA, RelB or c-Rel that contain transcriptional activation domains, these complexes engage with the basal transcriptional machinery to activate gene expression.

Signalling through the TNF-Receptor I induces the classical/canonical pathway through activation of the I kappa B kinase (IKK) complex formed by the regulatory subunit NEMO/IKKγ and the two catalytic subunits IKKα and IKKβ. IKK-mediated phosphorylation and subsequent degradation of the inhibitor IκBα results in nuclear translocation of the NF-κB1/RelA complex and activation of transcription (Fig. 4.2).

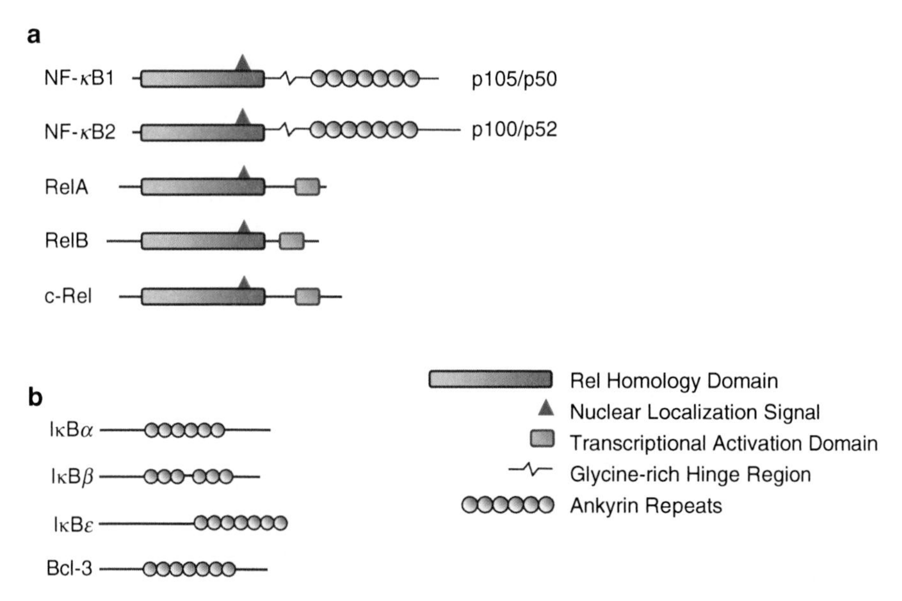


**Fig. 4.1** Structure of the NF-κB Proteins. (**a**) The NF-κB family of proteins in mammalian cells contains 5 members NF-κB1 p105/p50, NF-κB2 p100/p52, RelA, RelB and c-Rel. Common to all NF-κB family members is a region called Rel homology domain (RHD) that is located in the amino terminal part of these proteins. The RHD contains the DNA binding region, the dimerization domain and the nuclear localization signal. According to the structure, these proteins can be divided in two groups. The first group contains NF-κB1 (p105) and NF-κB2 (p100), that are synthesized as large precursors that upon cell stimulation and limited processing of the ankyrin repeats-containing regions by the proteasome generate the DNA binding subunits NF-κB1 p50 and NF-κB2 p52. Lack of transcriptional activation domains in p50 and p52 means that their homodimers will bind DNA and repress gene expression. RelA, RelB, and c-Rel contain transcriptional activation domains and form the second group. Heterodimers containing proteins of the second group will engage the basal transcriptional machinery to activate mRNA synthesis. (**b**) Structure of the IκB Proteins. The IκB proteins contain 6 or 7 ankyrin repeat motifs in their structure that bind to the RHD of the NF-κB proteins and thus prevent the latter from translocating to the cell nuclei to drive gene expression. Bcl-3 is an atypical IκB family protein since it binds to NF-κB2 homodimers to induce gene expression. Bcl-3 also binds to NF-κB1 homodimers

Signalling by the LTβR, RANK, CD40, TLR4 converge in the induction of the classical pathway as above but also the alternative/non-canonical pathway through activation of NF-κB-inducing kinase (NIK) and subsequent IKKα phosphorylation of the NF-κB2 p100 protein that upon limited proteasomal degradation of the carboxi terminus ankyrin repeats generates p52/RelB and p52/RelA heterodimers (Yamada et al. 2000; Lin et al. 2001; Xiao et al. 2001; Dejardin et al. 2002; Ghosh and Karin 2002; Mordmuller et al. 2003; Muller and Siebenlist 2003; Yilmaz et al. 2003; Dejardin 2006). The classical pathway induces the expression of the substrates of the alternative pathway, *Nfkb2* and *Relb*, therefore both pathways together could be considered as a single extended one that gets activated upon cell stimulation through specific receptors.

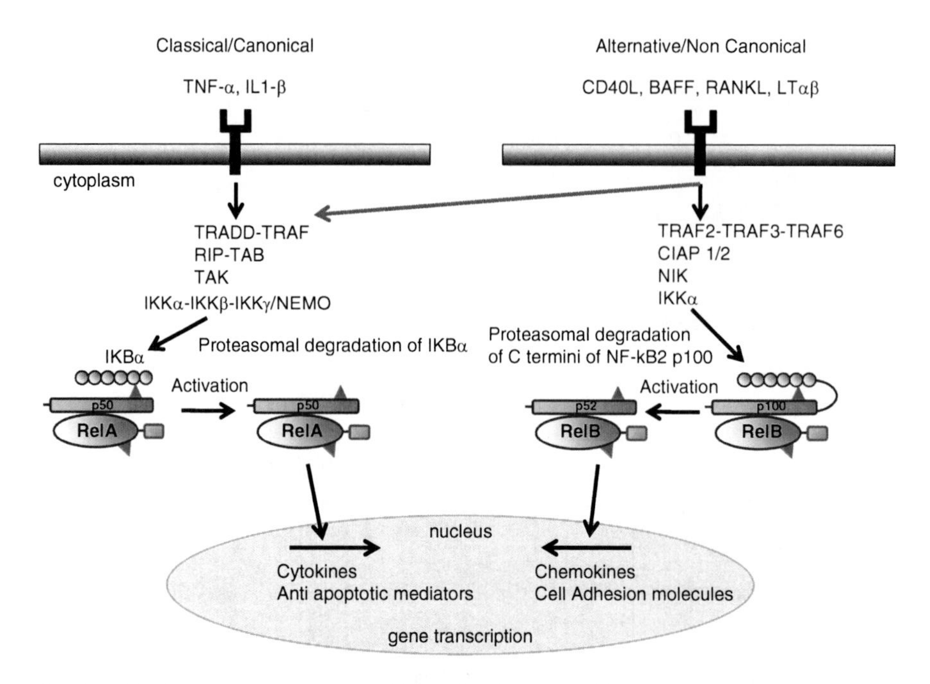


**Fig. 4.2** NF-κB Activation Pathways. The activation of the NF-κB proteins occurs by at least two different mechanisms. The classical/canonical pathway is activated by the TNF-Receptors or IL-1- Receptor and involves the activation of the IKK proteins, that through IKKβ will phosphorylate the IκBs to release the heterodimers from the cytoplasm and translocate to the cell nuclei. In contrast, the alternative/non canonical pathway through NIK and IKKa will phosphorylate the carboxi terminus of p100 and thus target the region for ubiquitination and parcial proteolisis by the proteosome. The latter process releases p52 containing complexes that translocate to the nuclei to activate transcription. Among the transcriptional targets of the classical pathway are IKBa, NF-κB2, RelB, cytokines and antiapoptotic mediators. Thus signalling by the classical pathway is required for the expression of the substrates of the alternative pathway. Among the transcriptional targets of the alternative pathway are homeostatic chemokines and cell adhesion molecules

Recent studies on signalling by the alternative pathway showed that NIK is essentially unstable in non-stimulated cells by being targeted for proteosome degradation by the ubiquitin ligases known as cellular inhibitors of apoptosis (cIAP) 1/2. The TNF receptor-associated factors, (TRAF) 2 and TRAF-3, act as a platform bringing together NIK and the cIAPs. Upon LTβR, BAFF-R or CD40 engagement, TRAF2 and 3 are degraded, NIK becomes stable and activates IKKα and subsequent NF-κB2 p100 processing (Qing et al. 2005; Vallabhapurapu et al. 2008; Sun and Ley 2008). In agreement with this mechanism, *Traf2*$^{-/-}$ and *Traf3*$^{-/-}$ mice present with a phenotype similar to mice carrying a constitutively active alternative NF-κB pathway, the *Nfkb2* p100Δ knock-in mice (p100Δ$^{-/-}$) (see below). Furthermore, the early lethality of the *Traf3*$^{-/-}$ mice can be rescued by crosses with the *Nfkb2*$^{-/-}$ mice (He et al. 2006).

Saccani and colleagues have shown a dimer exchange mechanism that regulates the expression of different NF-κB-target genes. Such mechanism is based on the fact that different dimers will bind sequentially to a specific kappa-B site and result in induction of gene transcription followed by repression or vice versa (Saccani et al. 2003). In some cases, dimer exchange might result in sustained expression of certain genes.

Recent evidence indicates that de-regulation of NF-κB proteins has an important role in inflammation and cancer (Karin and Greten 2005; Hagemann et al. 2007; Mantovani et al. 2008; Grivennikov et al. 2010). Interestingly, the development of SLO resembles chronic inflammation since interactions between bone marrow-derived cells and stromal cells are mediated by the same molecules during both processes (Kratz et al. 1996; Peduto et al. 2009). Based on these observations, SLO development can be seen as an inflammatory process in the sterile environment of embryogenesis.

## 4.3 LTβR Signalling and Activation of the NF-kB Pathways in Lymphoid Organ Development and Maintenance

The essential function of LTβR during SLO development is demonstrated by the absence of LNs, Peyer's patches (PP) and isolated lymphoid follicles (ILFs) in *Ltbr*$^{-/-}$ and *Lta*$^{-/-}$ mice, while *Ltb*$^{-/-}$ mice develop only mesenteric LNs (Rennert et al. 1996, 1997, 1998). One of the initial steps in SLO formation in embryos involves the crosstalk interactions between Lymphoid Tissue inducer (LTi) (CD45$^{+}$ CD4$^{+}$LTα1β2$^{+}$CXCR5$^{+}$RANK$^{+}$RANKL$^{+}$) cells and stromal cells (VCAM-1$^{+}$ICAM-1$^{+}$LTβR$^{+}$). LTi cells express LTα1β2 that upon engaging LTβR in stromal cells will induce a gene expression program through the NF-κB classical and alternative pathways (Mebius 2003; Cupedo and Mebius 2005; Drayton et al. 2006; Randall et al. 2008). This cellular crosstalk will result in the expression of Mucosal Cell Adhesion Molecule 1 (MAdCAM-1) and the homing chemokines CCL19, CCL21 and CXCL13 by stromal cells as well as contribute to their homeostasis (White et al. 2007). The expression of these and other unknown molecules facilitates further clustering of LTi cells with stromal organizer cells and the organization of specific B and T cell areas in the developing organs by generating a positive feedback loop. A similar impairment in expression of homing chemokines and cell adhesion molecules results in lack of B and T cell-specific areas in the spleen of *Ltbr*$^{-/-}$ mice (Ngo et al. 1999).

Studies of the different *Nfkb*-deficient mouse strains revealed the requirement and respective importance of the different component of the proteins of this signalling cascade activated upon engagement of LTβR (Weih and Caamaño 2003).

### 4.3.1 Relb-Deficient Mice

*Relb*$^{-/-}$ mice lack all LNs, PP and nasal-associated lymphoid tissues (NALT) due to a stromal cell defect resembling *Ltbr*$^{-/-}$ mice highlighting the absolute requirement of the alternative NF-κB pathway in the development of the SLO stroma (Weih et al. 1995; Weih et al. 2001; Yilmaz et al. 2003; Weih and Caamano 2003). *Relb* is not expressed in the mesenchymal precursor cells of the LN anlage and thus it does not seem to be necessary for the initial development of these structures in the mouse embryo, but its latter acquisition allows a strong expression of alternative pathway prototypic genes like *Madcam-1, Ccl19* and *Cxcl13* (Benezech et al. 2010).

Spleens of *Relb*$^{-/-}$ mice present defects in their micro-architecture, decreased expression of homing chemokines, a lack of MAdCAM-1 in the marginal sinus and absence of B cell follicles and germinal centres (GC) (Weih et al. 2001). P52/RelB heterodimers activate a specific set of genes and appear to be recruited to the promoters of *Ccl19* and *Ccl21* when activated by LTβR (Dejardin et al. 2002; Bonizzi and Karin 2004; Bonizzi et al. 2004).

### 4.3.2 Nfkb2-Deficient Mice and p100Δ Knock-In Mice

In contrast to *Relb*$^{-/-}$ mice, *Nfkb2*$^{-/-}$ mice exhibit only a limited phenotype when compared with the *Ltbr*$^{-/-}$ mice, presenting markedly small inguinal LNs and lacking popliteal LNs, PP and NALT due to a stromal cell defect (Paxian et al. 2002; Weih and Caamano 2003; Carragher et al. 2004). The development of high endothelial venules in inguinal LNs of the *Nfkb2*$^{-/-}$ mice is markedly impaired as shown upon blocking LTβR in vivo (Carragher et al. 2004; Browning et al. 2005). *Nfkb2*$^{-/-}$ spleens resemble the spleen of *RelB*$^{-/-}$ mice with their altered micro-architecture and lack of B cell follicles, GC and FDC (Caamaño et al. 1998; Franzoso et al. 1998; Poljak et al. 1999; Weih et al. 2001).

Constitutive signalling through the alternative NF-κB pathway, as occurs in the p100Δ$^{-/-}$ mice, leads to up-regulation of *Ccl21*, and the cell adhesion molecules *Madcam-1*, *Icam-1* and *Vcam-1* in the spleen (Ishikawa et al. 1997; Guo et al. 2007). Moreover, the micro-architecture of the spleen is disturbed with induction of ectopic HEV-like structures and absence of a normal marginal sinus (Guo et al. 2007). These observations argue that a tight regulation of the alternative NF-κB pathway is necessary to allow normal development of the spleen.

### 4.3.3 Nik-Deficient Mice, aly/aly Mice and IKKa$^{aa}$ Knock-In Mice

Animals deficient in the NF-κB-inducing kinase (*Nik and aly/aly*) lack all LNs and PPs, while kinase dead knock-in *Ikka*$^{aa}$ mice show a phenocopy of the *Nfkb2*$^{-/-}$

mice, having defects in inguinal and popliteal LNs (Yamada et al. 2000; Bonizzi et al. 2004; Drayton et al. 2004).

The fact that both *Nfkb2*$^{-/-}$ mice and *Ikka*$^{aa}$ knock-in mice exhibit similar milder defects in LN development when compared with *Rel*b$^{-/-}$ mice indicates that p50/RelB heterodimers, which are present in the former strains, partially compensate for the absence of p52/RelB heterodimers. Moreover, p50/RelB dimers have been shown to be regulated by NF-κB2 p100 (Derudder et al. 2003).

### 4.3.4 Rela-Deficient Mice

The embryonic lethal phenotype of the *Rela*$^{-/-}$ mice can be rescued by crosses with *Tnfr1*$^{-/-}$ or *Tnfa*$^{-/-}$ mice. The *Rela*$^{-/-}$/*Tnfr1*$^{-/-}$ mice are viable and lack all LNs due to a stromal cell defect implicating the role of RelA and the classical pathway in physiological LTβR signalling (Alcamo et al. 2001; Alcamo et al. 2002).

### 4.3.5 Nfkb1-Deficient Mice

The role of NF-κB1 in LTβR and RANK signalling during SLO development and organization is less clear than the function of other NF-κB family members. Interestingly, 75% of *Nfkb1*$^{-/-}$ mice lack inguinal LN, suggesting that NF-κB1 is required to a certain extent during the development of LN that also requires NF-κB2 (Lo et al. 2006).

### 4.3.6 Nfkb1/Nfkb2-Double Deficient Mice

Studies of *Nfkb1*$^{-/-}$ *Nfkb2*$^{-/-}$ double KO mice revealed an essential cooperation between NF-κB1 and NF-κB2 for fully functional LTβR signalling during SLO development. Indeed, while *Nfkb2*$^{-/-}$ mice present with a mild phenotype when compared with *Ltbr*$^{-/-}$ mice, the *Nfkb1/2*$^{-/-}$ double KO mice lack all LNs and exhibit a splenic microarchitecture that is markedly disorganized (Franzoso et al. 1998; Lo et al. 2006). Compound heterozygous mice for *Nfkb1 and Nfkb2* presented with a more marked phenotype that the single mutants suggesting a degree of functional redundancy between these two proteins during SLO development (Lo et al. 2006). In addition, activation of LTβR on MEFs induced the formation of the less well-characterized p52/RelA and p50/RelB dimers suggesting that those complexes also have distinct and important regulatory functions (Lovas et al. 2008).

## 4.4 TNFα-Induced Activation of NF-κB During SLO Development

The function of TNFα in SLO development is partly redundant with LTβR. There is mounting evidence that activation of the classical pathway through TNFα is necessary for the expression of some of the alternative NF-κB pathway target genes. For example, development of PP requires both LTβR and TNF-RI engagement as both *Ltbr*$^{-/-}$ mice and *Tnf-rI*$^{-/-}$ mice present with a defect in PP development. In particular, engagement of TNF-RI and activation of TRAF-2 allow the recruitment of RelA to the *Cxcl13* promoter and up-regulation of *Cxcl13* mRNA (Piao et al. 2007). Only the concomitant activation of TNF-RI and LTβR allowed a potent *Cxcl13* expression in LN stromal cells, which could not be achieved by an excess of RelB or p52 (Katakai et al. 2004; Suto et al. 2009).

The formation of aortic tertiary lymphoid organs during atherosclerosis has been shown recently to take place under the combine effect of TNF-RI and LTβR on smooth muscle cells to become lymphoid tissue organizer cells, resulting in an enhanced expression of homeostatic chemokines and recruitment of hematopoietic cells (Suto et al. 2009; Grabner et al. 2009). Furthermore, the fact that several LTβR-target genes on FDCs and germinal centre stromal cells have been shown to be down-regulated in spleens of both *Ltbr*$^{-/-}$ mice and *Tnf-rI*$^{-/-}$ mice gives support to the model of interconnection/cooperation between the TNF-RI-p50/RelA classical NF-κB inflammatory pathway and the LTβR-p52/RelB lymphorganogenic pathway (Huber et al. 2005; Basak et al. 2007; Basak and Hoffmann 2008; Basak et al. 2008; Lotzer et al. 2010). Likewise, a microarray analysis of LTβR activation on mouse embryo fibroblasts (MEF) deficient in *Rela* or *Relb* showed that expression of most LTβR-induced target genes require the presence of both NF-κB proteins (Lovas et al. 2008).

These results highlight the importance of the cross-talk between the classical and alternative NF-κB pathways and revealed the underestimated function of TNF-RI in supporting the development of SLO and tertiary LO.

## 4.5 RANK and Activation of NF-κB in Lymphoid Organ Development

The absence of most LN except for rudimentary mesenteric LN in *Rankl*$^{-/-}$ mice indicates that RANK-L and RANK are both important for SLO development. LTi cells express both RANK and RANK-L, while LN stromal cells express RANK-L. Interestingly, *Rankl*$^{-/-}$ mice present with reduced numbers of LTi cells indicating an important role of this pathway for their homeostasis (Dougall et al. 1999; Kong et al. 1999; Kim et al. 2000).

Yoshida and coworkers have shown that RANK signalling induces the expression of the LTβR ligand in LTi cells (Yoshida et al. 2002). Interestingly, LTβR signalling induces the expression of RANK-L in stromal cells (Vondenhoff et al. 2009). Concomitantly, mLNs of *Ltb*r$^{-/-}$ mice showed significant decreased expression of RANK-L (Benezech et al. 2010). Thus, a positive feedback loop between LTi cells and stromal organizer cells that enhances LTi cell clustering with organizer cells takes place.

Differentiation and homeostasis of LTi cells from their precursors also requires the transcription factors retinoid-like orphan receptor gamma RORγt and the transcriptional inhibitor helix-loop-helix protein Id2 (Yokota et al. 1999; Sun et al. 2000; Eberl et al. 2004). Recent studies have shown that RANK signalling in mammary epithelial cells triggers nuclear translocation of Id2 that is necessary for down-regulation of p21 and cell proliferation (Kim et al. 2006). Whether a similar mechanism links RANK and Id2 in LTi cells and whether NF-κB has a role in the homeostasis of these cells remain to be investigated.

## 4.6 NF-κB Function During Development and Maturation of Inducible Lymphoid Organs

An elegant report has demonstrated the reciprocal regulation between the bacterial population of the ileum and the development of ILFs in the intestine. Gram-negative bacteria induce the formation of ILFs. The transition between cryptopatches and mature ILFs requires signalling by the intracellular innate immune receptor NOD-1 while TLR2/4, TRIF, Myd88 and NOD-2 are necessary for full maturation of ILFs (Bouskra et al. 2008). Due to the involvement of TLR2/4, TRIF, Myd88 and NOD-2 in activation of the NF-κB pathways, it could be deduced that members of this family of transcription factors have an important role in this process.

## 4.7 Conclusion and Remaining Questions

The results discussed here show a clear function for the NF-κB proteins on stromal cells by contributing to their homeostasis and inducing the expression of chemokines and cell adhesion molecules that are essential for LN, PPs and ILFs development. Based on the analysis of the different *Nfkb* mutant mouse strains, a hierarchy for these factors during SLO development emerges with RelA and RelB been essential for LN development while NF-κB2 and NF-κB1 having a role in inducing the expression of genes that facilitate homing of lymphocytes to the LNs, PPs and ILFs. However, a series of questions remain still unanswered about the role of this family of proteins in both SLO development and organization as well as during inflammatory processes.

To what extent the absence of LNs in mice deficient in the classical/canonical NF-κB pathway (*Rela*$^{-/-}$/*Tnfr1*$^{-/-}$ mice) is due to the absence of the substrates of the alternative/non-canonical pathway (NF-κB2 and RelB) and the target genes of the latter?

What is the role of NF-κB in the homeostasis of the intestinal bacterial flora and the ILF formation? While several recent reports have shown that the IKK kinases have well defined cell-specific roles during intestinal inflammation and cancer, it still remains not fully defined the role of the individual NF-κB factors during these diseases (Nenci et al. 2007; Zaph et al. 2007; Pasparakis 2008; Pasparakis 2009).

Will it be possible to interfere with specific NF-κB proteins by means of cell-specific inhibitors in autoimmune diseases that present with tertiary lymphoid organs that perpetuate the symptoms?

Will it be possible through the manipulation of NF-κB activation to induce the formation of tertiary lymphoid organs in the proximity of specific tumours to facilitate the development of immune responses against them? (Carragher et al. 2008).

In other words, are the NF-κB proteins of the alternative pathway putative therapeutic targets for the treatment of autoimmune diseases and cancer?

Taking into account the essential role of LTβR-NIK-NF-κB signalling during SLO formation, will it be possible to use NIK inhibitors or agonists of c-IAP function to block activation of the alternative NF-κB pathway during chronic inflammatory diseases/formation of ectopic lymphoid tissues? (Gommerman and Browning 2003).

**Acknowledgments** This work was supported by a BBSRC project grant and a EU FP7 INFLA-CARE Collaborative Project grant to J. Caamaño and Deutsche Forschungsgemeinschaft (WE-2224/2, WE 2224/4, and WE-2224/5) grants to F. Weih.

## References

Alcamo E, Mizgerd JP, Horwitz BH, Bronson R, Beg AA, Scott M, Doerschuk CM, Hynes RO, Baltimore D (2001) Targeted mutation of TNF receptor. I rescues the RelA-deficient mouse and reveals a critical role for NF-kappa B in leukocyte recruitment. J Immunol 167: 1592–1600

Alcamo E, HacohenN, SchulteL, RennertP, HynesR, BaltimoreD (2002) Requirement of the NF-kB family member RelA in the development of secondary lymphoid organs. J Exp Med 195:233–244

Basak S, Hoffmann A (2008) Crosstalk via the NF-kappaB signaling system. Cytokine Growth Factor Rev 19:187–197

Basak S, Kim H, Kearns JD, Tergaonkar V, O'Dea E, Werner SL, Benedict CA, Ware CF, Ghosh G, Verma IM, Hoffmann A (2007) A fourth IkappaB protein within the NF-kappaB signaling module. Cell 128:369–381

Basak S, Shih VF, Hoffmann A (2008) Generation and activation of multiple dimeric transcription factors within the NF-kappaB signaling system. Mol Cell Biol 28:3139–3150

Benezech C, White A, Mader E, Serre K, Parnell S, Pfeffer K, Ware CF, Anderson G, Caamano JH (2010) Ontogeny of stromal organizer cells during lymph node development. J Immunol 184:4521–4530

Bonizzi G, Karin M (2004) The two NF-kappaB activation pathways and their role in innate and adaptive immunity. Trends Immunol 25:280–288

Bonizzi G, Bebien M, Otero DC, Johnson-Vroom KE, Cao Y, Vu D, Jegga AG, Aronow BJ, Ghosh G, Rickert RC, Karin M (2004) Activation of IKKalpha target genes depends on recognition of specific kappaB binding sites by RelB:p52 dimers. Embo J 23:4202–4210

Bouskra D, Brezillon C, Berard M, Werts C, Varona R, Boneca IG, Eberl G (2008) Lymphoid tissue genesis induced by commensals through NOD1 regulates intestinal homeostasis. Nature 456:507–510

Browning JL, Allaire N, Ngam-Ek A, Notidis E, Hunt J, Perrin S, Fava RA (2005) Lymphotoxin-beta receptor signaling is required for the homeostatic control of HEV differentiation and function. Immunity 23:539–550

Caamaño J, Rizzo C, Durham S, Barton D, Raventos-Suarez C, Snapper C, Bravo R (1998) Nuclear Factor (NF)-kB2 (p100/p52) is required for Normal Splenic Microarchitecture and B Cell-mediated Immune Responses. J Exp Med 187:185–196

Carragher D, Johal R, Button A, White A, Eliopoulos A, Jenkinson E, Anderson G, Caamano J (2004) A stroma-derived defect in NF-kappaB2$^{-/-}$ mice causes impaired lymph node development and lymphocyte recruitment. J Immunol 173:2271–2279

Carragher DM, Rangel-Moreno J, Randall TD (2008) Ectopic lymphoid tissues and local immunity. Semin Immunol 20:26–42

Cupedo T, Mebius RE (2005) Cellular interactions in lymph node development. J Immunol 174:21–25

Dejardin E (2006) The alternative NF-kappaB pathway from biochemistry to biology: pitfalls and promises for future drug development. Biochem Pharmacol 72:1161–1179

Dejardin E, Droin NM, Delhase M, Haas E, Cao Y, Makris C, Li ZW, Karin M, Ware CF, Green DR (2002) The lymphotoxin-beta receptor induces different patterns of gene expression via two NF-kappaB pathways. Immunity 17:525–535

Derudder E, Dejardin E, Pritchard LL, Green DR, Korner M, Baud V (2003) RelB/p50 dimers are differentially regulated by tumor necrosis factor-alpha and lymphotoxin-beta receptor activation: critical roles for p100. J Biol Chem 278:23278–23284

Dougall WC, Glaccum M, Charrier K, Rohrbach K, Brasel K, De Smedt T, Daro E, Smith J, Tometsko ME, Maliszewski CR, Armstrong A, Shen V, Bain S, Cosman D, Anderson D, Morrissey PJ, Peschon JJ, Schuh J (1999) RANK is essential for osteoclast and lymph node development. Genes Dev 13:2412–2424

Drayton DL, Bonizzi G, Ying X, Liao S, Karin M, Ruddle NH (2004) IkappaB kinase complex alpha kinase activity controls chemokine and high endothelial venule gene expression in lymph nodes and nasal-associated lymphoid tissue. J Immunol 173:6161–6168

Drayton DL, Liao S, Mounzer RH, Ruddle NH (2006) Lymphoid organ development: from ontogeny to neogenesis. Nat Immunol 7:344–353

Eberl G, Marmon S, Sunshine MJ, Rennert PD, Choi Y, Littman DR (2004) An essential function for the nuclear receptor RORgamma(t) in the generation of fetal lymphoid tissue inducer cells. Nat Immunol 5:64–73

Franzoso G, Carlson L, Poljak L, Shores E, Epstein S, Leonardi A, Grinberg A, Tran T, Scharton-Kersten T, Anver M, Love P, Brown K, Siebenlist U (1998) Mice deficient in nuclear factor (NF)-kB/p52 present with defects in humoral responses, germinal center reactions, and splenic microarchitecture. J Exp Med 187:147–159

Fu Y-X, Chaplin D (1999) Development and maturation of secondary lymphoid tissues. Annu Rev Immunol 17:399–433

Ghosh S, Hayden MS (2008) New regulators of NF-kappaB in inflammation. Nat Rev Immunol 8:837–848

Ghosh S, Karin M (2002) Missing pieces in the NF-kappaB puzzle. Cell 109 Suppl:S81–S96

Gommerman JL, Browning JL (2003) Lymphotoxin/light, lymphoid microenvironments and autoimmune disease. Nat Rev Immunol 3:642–655

Grabner R, Lotzer K, Dopping S, Hildner M, Radke D, Beer M, Spanbroek R, Lippert B, Reardon CA, Getz GS, Fu YX, Hehlgans T, Mebius RE, van der Wall M, Kruspe D, Englert C, Lovas A, Hu D, Randolph GJ, Weih F, Habenicht AJ (2009) Lymphotoxin beta receptor signaling

promotes tertiary lymphoid organogenesis in the aorta adventitia of aged ApoE$^{-/-}$ mice. J Exp Med 206:233–248
Grivennikov SI, Greten FR, Karin M (2010) Immunity, inflammation, and cancer. Cell 140:883–899
Guo F, Weih D, Meier E, Weih F (2007) Constitutive alternative NF-kappaB signaling promotes marginal zone B-cell development but disrupts the marginal sinus and induces HEV-like structures in the spleen. Blood 110:2381–2389
Hagemann T, Balkwill F, Lawrence T (2007) Inflammation and cancer: a double-edged sword. Cancer Cell 12:300–301
He JQ, Zarnegar B, Oganesyan G, Saha SK, Yamazaki S, Doyle SE, Dempsey PW, Cheng G (2006) Rescue of TRAF3-null mice by p100 NF-kappa B deficiency. J Exp Med 203:2413–2418
Huber C, Thielen C, Seeger H, Schwarz P, Montrasio F, Wilson MR, Heinen E, Fu YX, Miele G, Aguzzi A (2005) Lymphotoxin-beta receptor-dependent genes in lymph node and follicular dendritic cell transcriptomes. J Immunol 174:5526–5536
Ishikawa H, Carrasco D, Claudio E, Ryseck RP, Bravo R (1997) Gastric hyperplasia and increased proliferative responses of lymphocytes in mice lacking the COOH-terminal ankyrin domain of NF-kappaB2. J Exp Med 186:999–1014
Karin M, Greten FR (2005) NF-kappaB: linking inflammation and immunity to cancer development and progression. Nat Rev Immunol 5:749–759
Katakai T, Hara T, Sugai M, Gonda H, Shimizu A (2004) Lymph node fibroblastic reticular cells construct the stromal reticulum via contact with lymphocytes. J Exp Med 200:783–795
Kim D, Mebius RE, MacMicking JD, Jung S, Cupedo T, Castellanos Y, Rho J, Wong BR, Josien R, Kim N, Rennert PD, Choi Y (2000) Regulation of peripheral lymph node genesis by the tumor necrosis factor family member TRANCE. J Exp Med 192:1467–1478
Kim NS, Kim HJ, Koo BK, Kwon MC, Kim YW, Cho Y, Yokota Y, Penninger JM, Kong YY (2006) Receptor activator of NF-kappaB ligand regulates the proliferation of mammary epithelial cells via Id2. Mol Cell Biol 26:1002–1013
Kong YY, Yoshida H, Sarosi I, Tan HL, Timms E, Capparelli C, Morony S, Oliveira-dos-Santos AJ, Van G, Itie A, Khoo W, Wakeham A, Dunstan CR, Lacey DL, Mak TW, Boyle WJ, Penninger JM (1999) OPGL is a key regulator of osteoclastogenesis, lymphocyte development and lymph-node organogenesis. Nature 397:315–323
Kratz A, Campos-Neto A, Hanson MS, Ruddle NH (1996) Chronic inflammation caused by lymphotoxin is lymphoid neogenesis. J Exp Med 183:1461–1472
Lin Y, Wu L, Wesche H, Arthur C, White J, Goeddel D, Schreiber R (2001) Defective Lymphotoxin-b-Receptor-induced NF-kB transcriptional activity in NIK-deficient mice. Science 291:2162–2165
Lo JC, Basak S, James ES, Quiambo RS, Kinsella MC, Alegre ML, Weih F, Franzoso G, Hoffmann A, Fu YX (2006) Coordination between NF-kappaB family members p50 and p52 is essential for mediating LTbetaR signals in the development and organization of secondary lymphoid tissues. Blood 107:1048–1055
Lotzer K, Dopping S, Connert S, Grabner R, Spanbroek R, Lemser B, Beer M, Hildner M, Hehlgans T, van der Wall M, Mebius RE, Lovas A, Randolph GJ, Weih F, Habenicht AJ (2010) Mouse aorta smooth muscle cells differentiate into lymphoid tissue organizer-like cells on combined tumor necrosis factor receptor- 1/lymphotoxin beta-receptor NF-kappaB signaling. Arterioscler Thromb Vasc Biol 30:395–402
Lovas A, Radke D, Albrecht D, Yilmaz ZB, Moller U, Habenicht AJ, Weih F (2008) Differential RelA- and RelB-dependent gene transcription in LTbetaR- stimulated mouse embryonic fibroblasts. BMC Genomics 9:606
Mantovani A, Allavena P, Sica A, Balkwill F (2008) Cancer-related inflammation. Nature 454:436–444
Mebius RE (2003) Organogenesis of lymphoid tissues. Nat Rev Immunol 3:292–303

Mordmuller B, Krappmann D, Esen M, Wegener E, Scheidereit C (2003) Lymphotoxin and lipopolysaccharide induce NF-kappaB-p52 generation by a co-translational mechanism. EMBO Rep 4:82–87

Mueller SN, Germain RN (2009) Stromal cell contributions to the homeostasis and functionality of the immune system. Nat Rev Immunol 9:618–629

Muller JR, Siebenlist U (2003) Lymphotoxin beta receptor induces sequential activation of distinct NF-kappa B factors via separate signaling pathways. J Biol Chem 278:12006–12012

Nenci A, Becker C, Wullaert A, Gareus R, van Loo G, Danese S, Huth M, Nikolaev A, Neufert C, Madison B, Gumucio D, Neurath MF, Pasparakis M (2007) Epithelial NEMO links innate immunity to chronic intestinal inflammation. Nature 446:557–561

Ngo V, Korner H, Gunn M, Schmidt K, Riminton D, Cooper M, Browning J, Sedgwick J, Cyster J (1999) Lymphotoxin a/b and tumor necrosis factor are required for stromal cell expression of homing chemokines in B and T cell areas of the spleen. J Exp Med 189:403–412

Pasparakis M (2008) IKK/NF-kappaB signaling in intestinal epithelial cells controls immune homeostasis in the gut. Mucosal Immunol 1 (Suppl 1):S54–S57

Pasparakis M (2009) Regulation of tissue homeostasis by NF-kappaB signalling: implications for inflammatory diseases. Nat Rev Immunol 9:778–788

Paxian S, Merkle H, Riemann M, Wilda M, Adler G, Hameister H, Liptay S, Pfeffer K, Schmid RM (2002) Abnormal organogenesis of Peyer's patches in mice deficient for NF-kappaB1, NF-kappaB2, and Bcl-3. Gastroenterology 122:1853–1868

Peduto L, Dulauroy S, Lochner M, Spath GF, Morales MA, Cumano A, Eberl G (2009) Inflammation recapitulates the ontogeny of lymphoid stromal cells. J Immunol 182:5789–5799

Perkins ND (2007) Integrating cell-signalling pathways with NF-kappaB and IKK function. Nat Rev Mol Cell Biol 8:49–62

Piao JH, Yoshida H, Yeh WC, Doi T, Xue X, Yagita H, Okumura K, Nakano H (2007) TNF receptor-associated factor 2-dependent canonical pathway is crucial for the development of Peyer's patches. J Immunol 178:2272–2277

Poljak L, Carlson L, Cunningham K, Kosco-Vilbois M, Siebenlist U (1999) Distinct activities of p52/NF-kB required for proper secondary lymphoid organ microarchitecture: functions enhanced by Bcl-3. J Immunol 163:6581–6588

Qing G, Qu Z, Xiao G (2005) Stabilization of basally translated NF-kappaB-inducing kinase (NIK) protein functions as a molecular switch of processing of NF-kappaB2 p100. J Biol Chem 280:40578–40582

Randall TD, Carragher DM, Rangel-Moreno J (2008) Development of secondary lymphoid organs. Annu Rev Immunol 26:627–650

Rennert PD, Browning JL, Mebius R, Mackay F, Hochman PS (1996) Surface lymphotoxin alpha/beta complex is required for the development of peripheral lymphoid organs. J Exp Med 184:1999–2006

Rennert P, Browning J, Hochman P (1997) Selective disruption of lymphotoxin ligands reveals a novel set of mucosal lymph nodes and unique effects on lymph node cellular organization. Int Immunol 9:1627–1639

Rennert P, James D, Mackay F, Browning J, Hochman P (1998) Lymph node genesis is induced by signaling through the Lymphotoxin b Receptor. Immunity 9:71–79

Saccani S, Pantano S, Natoli G (2003) Modulation of NF-kappaB activity by exchange of dimers. Mol Cell 11:1563–1574

Sun SC, Ley SC (2008) New insights into NF-kappaB regulation and function. Trends Immunol 29:469–478

Sun Z, Unutmaz D, Zou Y-R, Sunshine M, Pierani A, Brenner-Morton S, Mebius R, Littman D (2000) Requirement for RORa in thymocyte survival and lymphocyte organ development. Science 288:2369–2372

Suto H, Katakai T, Sugai M, Kinashi T, Shimizu A (2009) CXCL13 production by an established lymph node stromal cell line via lymphotoxin-beta receptor engagement involves the cooperation of multiple signaling pathways. Int Immunol 21:467–476

Vallabhapurapu S, Karin M (2009) Regulation and function of NF-kappaB transcription factors in the immune system. Annu Rev Immunol 27:693–733

Vallabhapurapu S, Matsuzawa A, Zhang W, Tseng PH, Keats JJ, Wang H, Vignali DA, Bergsagel PL, Karin M (2008) Nonredundant and complementary functions of TRAF2 and TRAF3 in a ubiquitination cascade that activates NIK-dependent alternative NF-kappaB signaling. Nat Immunol 9:1364–1370

Vondenhoff MF, Greuter M, Goverse G, Elewaut D, Dewint P, Ware CF, Hoorweg K, Kraal G, Mebius RE (2009) LTbetaR signaling induces cytokine expression and up-regulates lymphangiogenic factors in lymph node anlagen. J Immunol 182:5439–5445

Weih F, Caamaño J (2003) Regulation of secondary lymphoid organ development by the NF-kB signal transduction pathway. Imm Rev 195:91–105

Weih F, Carrasco D, Durham S, Barton D, Rizzo C, Ryseck R, Lira S, Bravo R (1995) Multiorgan inflammation and hematopoietic abnormalities in mice with a targeted disruption of RelB, a member of the NF-kB/Rel family. Cell 80:331–340

Weih DS, Yilmaz ZB, Weih F (2001) Essential role of RelB in germinal center and marginal zone formation and proper expression of homing chemokines. J Immunol 167:1909–1919

White A, Carragher D, Parnell S, Msaki A, Perkins N, Lane P, Jenkinson E, Anderson G, Caamano JH (2007) Lymphotoxin a-dependent and -independent signals regulate stromal organizer cell homeostasis during lymph node organogenesis. Blood 110:1950–1959

Xiao G, Harhaj E, Sun S-C (2001) NF-kB-inducing kinase regulates the processing of NF-kB2 p100. Mol Cell 7:401–409

Yamada T, Mitani T, Yorita K, Uchida D, Matsushima A, Iwamasa K, Fujita S, Matsumoto M 2000 Abnormal immune function of hemopoietic cells from alymphoplasia (aly) mice, a natural strain with mutant NF-kB-inducing kinase. J Immunol 165:804–812

Yilmaz ZB, Weih DS, Sivakumar V, WeihF (2003) RelB is required for Peyer's patch development: differential regulation of p52-RelB by lymphotoxin and TNF. Embo J 22:121–130

Yokota Y, Mansouri A, Mori S, Sugawara S, Adachi S, Nishikawa S, Gruss P (1999) Development of peripheral lymphoid organs and natural killer cells depends on the helix-loop-helix inhibitor Id2. Nature 397:702–706

Yoshida H, Naito A, Inoue J, Satoh M, Santee-Cooper SM, Ware CF, Togawa A, Nishikawa S (2002) Different cytokines induce surface lymphotoxin-alphabeta on IL-7 receptor-alpha cells that differentially engender lymph nodes and Peyer's patches. Immunity 17:823–833

Zaph C, Troy AE, Taylor BC, Berman-Booty LD, Guild KJ, Du Y, Yost EA, Gruber AD, May MJ, Greten FR, Eckmann L, Karin M, Artis D (2007) Epithelial-cell-intrinsic IKK-beta expression regulates intestinal immune homeostasis. Nature 446:552–556

# Chapter 5
# Homeostatic Chemokines, Cytokines and Their Receptors in Peripheral Lymphoid Organ Development

**Péter Balogh**

**Abstract** The tissue structure of secondary lymphoid organs is characterized by a highly compartmentalized distribution of its haematopoietic and stromal cells. The ordered arrangement of mobile lymphoid cells is assisted by the sessile stromal tissue constituents, whose cytokine production provides survival stimuli for lymphocytes as well as positioning cues within the various lymphocyte compartments. In this part, the developmental aspects of essential cytokines and homeostatic chemokines that also profoundly affect the embryonic formation of secondary lymphoid tissues are presented. Importantly, the same soluble factors may operate during the initiation as well as the maintenance of organized lymphoid structure, although the interacting haemopoietic and stromal cellular partners are substantially different between the two conditions. This part describes the main soluble factors and their combined effects on the development of peripheral lymphoid tissues.

## 5.1 IL-7 and Its Receptor: From Lymphocyte Differentiation to Lymphoid Organogenesis

The formation of committed hemopoietic lineages is the result of the concerted action of endogeneous programming and the environmental influences, including a variety of soluble factors. One of the earliest such molecules identified through its profound effects on the development of secondary lymphoid organs was interleukin 7 (IL-7). Although originally identified more than 20 years ago as an important growth factor for the differentiation of mouse pre-B cells in bone marrow (Namen et al. 1988) and developing thymocytes (Murray et al. 1989), and also for T cells in human (Noguchi et al. 1993), more recent observations have also established that its roles in both lymphoid organogenesis and maintenance of tissue capacity for

P. Balogh
Department of Immunology and Biotechnology, Faculty of Medicine, University of Pécs, Pécs, Hungary

P. Balogh (ed.), *Developmental Biology of Peripheral Lymphoid Organs*,
DOI 10.1007/978-3-642-14429-5_5, 

hosting immune responses go beyond than merely providing survival signals for those cells' precursors that fill up the secondary lymphoid organs.

IL-7 binds to the IL-7 receptor complex (IL-7R), consisting of an IL7R$\alpha$ chain (CD127) non-covalently associated with a gamma-chain ($\gamma$c). Of these two components, IL-7R$\alpha$ is shared with another receptor recognizing thymic stromal lymphopoietin (TSLP, Sims et al. 2000), an IL-7-like soluble protein produced by thymic, bronchial and intestinal epithelial cells, and is associated with its specific receptor subunit, TSLPR. On the other hand, $\gamma$c is shared with the receptor for other cytokines such as IL-2, IL-9, IL15 and IL-21, and these ligands exert a broad range of effects (survival, costimulation and expansion), primarily affecting T cell subsets (Rochman et al. 2009; Overwijk and Schluns 2009). TSLP and IL-7 may play complementary roles, as the embryonic B cell is largely IL-7-independent, although fetal B cells are TSLP-responsive; in addition, TSLP may also influence B-1 subset distribution (Astrakhan et al. 2007), while the B cell formation in adult bone marrow is IL-7 dependent (Vosshenrich et al. 2003; Dias et al. 2005). More recently TSLP production was also observed in synovial fibroblasts from patients with rheumatoid arthritis, which may indicate a role for this cytokine in lymphoid neogenesis associated with chronic inflammation (Ozawa et al. 2007), whereas the absence of TSLP signalling in *tslpr*$^{-/-}$ mice results in enhanced intestinal inflammatory response (Taylor et al. 2009). Compared with TSLP, however, considerably more information is available concerning the role of IL-7 and its receptor signalling in the formation of peripheral lymphoid organs.

## 5.2 Role of IL-7R Signalling in Lymphoid Organogenesis

The ligand binding by IL-7R induces an extensive array of cellular responses mediated by the Jak-Stat signalling pathway, comprising a highly complex machinery which is also shared by other cytokines. Jak1 is associated with the $\alpha$ chain of IL-4R and possibly IL-7R, whereas $\gamma$c is coupled to Jak3. Sequential tyrosine phosphorylation of the receptors themselves and the Jak components will generate SH2 docking sites, where Stat transcription factors may bind, and also launches PI3K-Akt pathway, respectively (Schindler et al. 2007). Illustrating the complexity of signalling, several cytokines and growth factors may activate seven Stat family members, where their effects can be modulated by the members of activated suppressors of cytokine signalling family (SOCSs). Of these, SOCS1 is the main modulator of IL-7 signalling in vivo (Fujimoto et al. 2000). As the ligand specificity of IL-7R is determined by the combination of IL-7R7alpha; and the $\gamma$c, the consequences of either the absence of IL-7 ligand or the specific receptor subunit IL-7R$\alpha$ paired with the downstream signalling mediated by $\gamma$c-associated Jak3 should indicate the involvement of IL-7 recognition in the formation of lymphoid tissues.

Initial observations on the formation of Peyer's patches reported a puzzling difference between the complete absence of PPs in mice with deleted IL-7R$\gamma$ (despite the presence of some mature lymphocytes), and the presence of initial

VCAM-1-clustering in rag2-deficient, mice, with no mature lymphocytes (Adachi et al. 1998). Subsequently, the injection of anti-IL7R$\alpha$ monoclonal antibody into pregnant mice was sufficient to block the development of PPs (Yoshida et al. 1999). The signalling function of Jak3 associated with $\gamma$c in promoting IL-7R-associated lymphoid organogenesis also showed that the absence of either the receptor or the adaptor/kinase Jak3 blocks the development of lymph nodes and PPs (Cao et al. 1995; Park et al. 1995). However, the absence of IL-7R$\alpha$ resulted in a variable deficiency of pLNs, including either their absence at certain anatomical location or their reduced size (Luther et al. 2003). Importantly, none of these deficiencies affected the development of spleen. However, due to a severe blockade of lymphocyte development in the thymus and bone marrow, the cellularity was markedly reduced, albeit gradually recovered. The absence of other peripheral lymphoid organs reinforces that the time window for the initiation of (most) secondary lymphoid organs is restricted, and the presence of mature lymphoid cells cannot compensate for the absence of crucial developmental stimuli at a preceding period.

What is the role of IL-7 itself in this scenario? Surprisingly, in mice with ablated IL-7, the blockade of pLN formation was variable. Lymph nodes (including mesenteric LNs) in some regions were present at the same frequency as in wild-type controls. In cervical, popliteal and visceral (renal, hepatic, pancreatic and paraaortic) locations, however, their numbers were drastically reduced (Meier et al. 2007) or absent, together with the Peyer's patches (von Freeden-Jeffry et al. 1995), thus broadening the complexity of the regulation of lymphoid organogenesis, even within the same type of such tissues. Currently, there are no comprehensive reports on the effect of TSLP deficiency on the formation of pLNs and PPs, although the presence of mLNs was noted in these mutants (Taylor et al. 2009). It remains to be seen whether the combined absence of IL-7 and TSLP as possible ligands for receptors with IL-7R$\alpha$ component can further expand the blockade of pLNs. On the other hand, augmented production of IL-7 induced the appearance of ectopic lymph nodes and follicles through increased number of LTi cells, and the addition of IL-7 increased the survival of LTi cells in co-culture with stromal cells, as well as their cell surface expression of LT$\alpha\beta$2 heterotrimer (Meier et al. 2007).

## 5.3 Homeostatic Chemokines and Receptors: Relay of LTi Cells and Mature Lymphocytes in Establishing and Maintaining Compartmentalized Lymphoid Tissue Architecture

A common hallmark of secondary lymphoid tissues is the ordered segregation of its lymphoid as well as their stromal constituents into discrete anatomical territories, divided into T and B cell zones corresponding to the majority lymphocyte population. A major breakthrough in our understanding the mechanisms responsible for this arrangement was the identification of the G-protein-linked chemokine receptor CXCR5 (originally denoted as Burkitt's lymphoma receptor-1/BLR1) as a crucial B cell-associated cell surface molecule involved in the proper positioning of B cells and a smaller CD4 T cell subset into the follicles (Förster et al. 1996). Interestingly, in addition to the aberrant segregation of lymphocytes in the spleen of mice with the absence of this molecule, the inguinal lymph nodes were also absent, with variable defects of PPs, whereas the other pLNs investigated were similar to the corresponding wild-type controls. Subsequent work identified CXCL13 (originally named B-lymphocyte chemoattractant – Blc) as CXCR5 ligand (Gunn et al. 1998). Furthermore, binding of CXCL13 by CXCR5 to B cells induces the upregulation of B cell-associated LTαβ2 in spleen. In lymph nodes, however, the cells expressing LTαβ2 in pLNs are different, and also the initial seeding and positioning of B cells shortly after birth were demonstrated to be CXCL13-independent (Cupedo et al. 2004). The mature structure in pLNs is established when B cells provide sufficient amount of LTαβ2 for inducing FDCs which, in turn, will secrete larger amount of CXCL13. In PPs, where IL-7Rα is crucially involved in inducing the upregulation of LTi-associated LTαβ2, CXCL13 may also facilitate the association between VCAM-1 ligand α4β7 integrin (Finke et al. 2002). Thus it appears that, depending on the tissue, several signalling mechanisms between varying cellular elements (LTi or mature lymphocytes with stromal cells and FDCs, respectively) converge to achieve a threshold signalling by LTαβ2, where the spectrum of signalling events also involves ligand recognition by IL-7Rα and CXCL13.

In addition to CXCL13/CXCR5 interaction, CCR7 chemokine receptor and its ligands, CCL21 and CCL19, respectively, have also been suggested to be involved in lymphoid organogenesis. In adult mice, recognition of CCL21 and CCL19 chemokines by CCR7 directs the migration of T cells into the lymph nodes via the high endothelial venules and DC via afferent lymphatics, respectively (Förster et al. 2008). Fetal LTi cells expressed both CXCR5 and CCR7 and migrated efficiently along CXCL13 and CCL19/CCL21 gradients in vitro (Honda et al. 2001). In addition, when CXCL13-deficiency was combined with the absence of CCR7, these multi-chemokine signal-defective mice showed an increased blockade of pLN formation – yet, mesenteric LNs still developed. Similarly, the absence of CXCR5 (which permitted the formation of some pLNs) combined with the deficiency of CCR7 (which had a normal set of pLNs) resulted in the complete lack of pLNs, although mLNs still developed (Ohl et al. 2003). However, the combination of CXCL13 deficiency with the absence of IL-7Rα led to the complete blockade of mesenteric LN formation, coupled with the lack of LTi cells, indicating that the survival of these cells crucially depends on the simultaneous signalling via CXCR5 and IL-7R (Luther et al. 2003). In addition, the absence of pLNs at identical anatomical regions (i.e. cervical or inguinal) was also highly variable in mice with either the single or combined deficiencies of these morphogenic elements. Furthermore, there is a clear discrepancy between the consequences of the combined absences of CXCL13 and CCL21/19 where some pLNs can still develop, and the lack of the CXCR5/CCR7 receptors for the above chemokines, leading to the absence of any detectable pLNs (Müller et al. 2003).Collectively, these observations establish that peripheral LNs in various anatomic locations and mLNs have a rather complex individual pattern of CXCL13/CXCR5, IL-7Rα and CCL21/CCR7

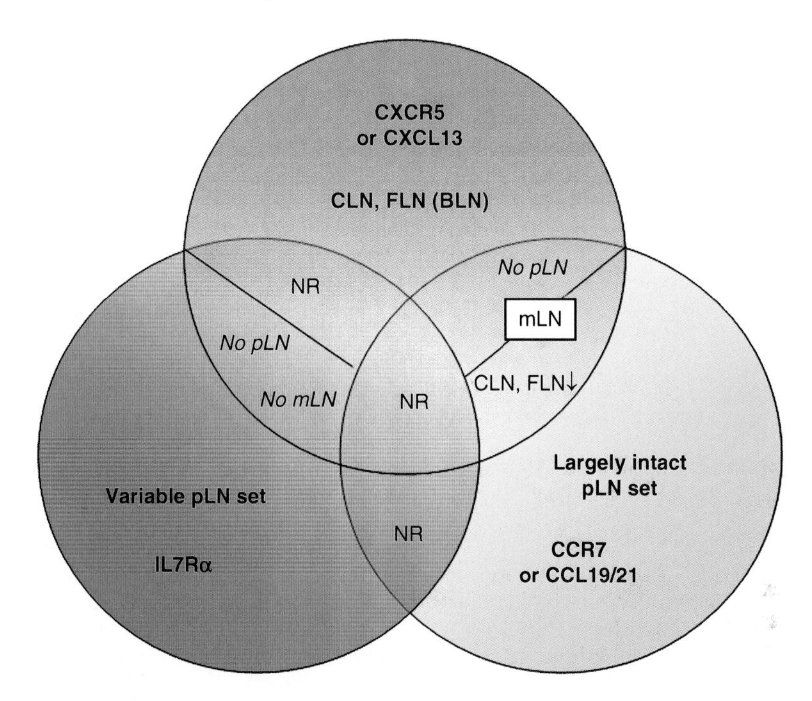


**Fig. 5.1** The results of the absence of various cytokines or their receptors on the development of lymph nodes. Cytokines and their receptors within each circle are indicated in *bold*, together with the lymph nodes formed (or their lack – in *italic*) in the absence of these molecules. Overlapping areas correspond to the combined phenotype the double mutants. The *upper half* segments of the areas shared between CXCR5-CXCL13/IL7Rα and CXCR5-CXCL13/CCR7/CCL19/21 correspond to the receptor–receptor double KO mutants, the *lower half* to the ligand–ligand 2x KO mutants, respectively, with mesenteric lymph nodes (mLNs) formed in both conditions. CLN, FLN and BLN are abbreviations for cervical, facial and brachial LNs, respectively; *NR* not reported

dependence for their formation. The kaleidoscope of the mutations observed in various combinations of these morphogenic factors is shown in Fig. 5.1.

## References

Adachi S, Yoshida H, Honda K, Maki K, Saijo K, Ikuta K, Saito T, Nishikawa SI (1998) Essential role of IL-7 receptor alpha in the formation of Peyer's patch anlage. Int Immunol 10(1):1–6

Astrakhan A, Omori M, Nguyen T, Becker-Herman S, Iseki M, Aye T, Hudkins K, Dooley J, Farr A, Alpers CE, Ziegler SF, Rawlings DJ (2007) Local increase in thymic stromal lymphopoietin induces systemic alterations in B cell development. Nat Immunol 8:522–531

Cao X, Shores EW, Hu-Li J, Anver MR, Kelsall BL, Russell SM, Drago J, Noguchi M, Grinberg A, Bloom ET et al (1995) Defective lymphoid development in mice lacking expression of the common cytokine receptor gamma chain. Immunity 2:223–238

Cupedo T, Lund FE, Ngo VN, Randall TD, Jansen W, Greuter MJ, de Waal-Malefyt R, Kraal G, Cyster JG, Mebius RE (2004) Initiation of cellular organization in lymph nodes is regulated by non-B cell-derived signals and is not dependent on CXC chemokine ligand 13. J Immunol 173:4889–4896

Dias S, Silva H Jr, Cumano A, Vieira P (2005) Interleukin-7 is necessary to maintain the B cell potential in common lymphoid progenitors. J Exp Med 201:971–979

Finke D, Acha-Orbea H, Mattis A, Lipp M, Kraehenbuhl J (2002) CD4(+)CD3(−) cells induce Peyer's patch development. Role of alpha4beta1 integrin activation by CXCR5. Immunity 17:363–373

Förster R, Mattis AE, Kremmer E, Wolf E, Brem G, Lipp M (1996) A putative chemokine receptor, BLR1, directs B cell migration to defined lymphoid organs and specific anatomic compartments of the spleen. Cell 87:1037–1047

Förster R, Davalos-Misslitz AC, Rot A (2008) CCR7 and its ligands: balancing immunity and tolerance. Nat Rev Immunol 8:362–371

Fujimoto M, Naka T, Nakagawa R, Kawazoe Y, Morita Y, Tateishi A, Okumura K, Narazaki M, Kishimoto T (2000) Defective thymocyte development and perturbed homeostasis of T cells in STAT-induced STAT inhibitor-1/suppressors of cytokine signaling-1 transgenic mice. J Immunol 165:1799–1806

Gunn MD, Ngo VN, Ansel KM, Ekland EH, Cyster JG, Williams LT (1998) A B-cell-homing chemokine made in lymphoid follicles activates Burkitt's lymphoma receptor-1. Nature 391:799–803

Honda K, Nakano H, Yoshida H, Nishikawa S, Rennert P, Ikuta K, Tamechika M, Yamaguchi K, Fukumoto T, Chiba T et al (2001) Molecular basis for hematopoietic/mesenchymal interaction during initiation of Peyer's patch organogenesis. J Exp Med 193:621–630

Luther SA, Ansel KM, Cyster JG (2003) Overlapping roles of CXCL13, interleukin 7 receptor alpha, and CCR7 ligands in lymph node development. J Exp Med 197:1191–1198

Meier D, Bornmann C, Chappaz S, Schmutz S, Otten LA, Ceredig R, Acha-Orbea H, Finke D (2007) Ectopic lymphoid-organ development occurs through interleukin 7-mediated enhanced survival of lymphoid-tissue-inducer cells. Immunity 26:643–654

Murray R, Suda T, Wrighton N, Lee F, Zlotnik A (1989) IL-7 is a growth and maintenance factor for mature and immature thymocyte subsets. Int Immunol 1:526–531

Müller G, Höpken UE, Lipp M (2003) The impact of CCR7 and CXCR5 on lymphoid organ development and systemic immunity. Immunol Rev 195:117–135

Namen AE, Lupton S, Hjerrild K, Wignall J, Mochizuki DY, Schmierer A, Mosley B, March CJ, Urdal D, Gillis S (1988) Stimulation of B-cell progenitors by cloned murine interleukin-7. Nature 333:571–573

Noguchi M, Nakamura Y, Russell SM, Ziegler SF, Tsang M, Cao X, Leonard WJ (1993) Interleukin-2 receptor gamma chain: a functional component of the interleukin-7 receptor. Science 262:1877–1880

Ohl L, Henning G, Krautwald S, Lipp M, Hardtke S, Bernhardt G, Pabst O, Förster R (2003) Cooperating mechanisms of CXCR5 and CCR7 in development and organization of secondary lymphoid organs. J Exp Med 197:1199–1204

Overwijk WW, Schluns KS (2009) Functions of gammaC cytokines in immune homeostasis: current and potential clinical applications. Clin Immunol 132:153–165

Ozawa T, Koyama K, Ando T, Ohnuma Y, Hatsushika K, Ohba T, Sugiyama H, Hamada Y, Ogawa H, Okumura K, Nakao A (2007) Thymic stromal lymphopoietin secretion of synovial fibroblasts is positively and negatively regulated by Toll-like receptors/nuclear factor-kappaB pathway and interferon-gamma/dexamethasone. Mod Rheumatol 17:459–463

Park SY, Saijo K, Takahashi T, Osawa M, Arase H, Hirayama N, Miyake K, Nakauchi H, Shirasawa T, Saito T (1995) Developmental defects of lymphoid cells in Jak3 kinase-deficient mice. Immunity 3:771–782

Rochman Y, Spolski R, Leonard WJ (2009) New insights into the regulation of T cells by gamma (c) family cytokines. Nat Rev Immunol 9:480–490

Schindler C, Levy DE, Decker T (2007) JAK-STAT signaling: from interferons to cytokines. J Biol Chem 282:20059–20063

Sims JE, Williams DE, Morrissey PJ, Garka K, Foxworthe D, Price V, Friend SL, Farr A, Bedell MA, Jenkins NA, Copeland NG, Grabstein K, Paxton RJ (2000) Molecular cloning and biological characterization of a novel murine lymphoid growth factor. J Exp Med 192:671–180
Taylor BC, Zaph C, Troy AE, Du Y, Guild KJ, Comeau MR, Artis D (2009) TSLP regulates intestinal immunity and inflammation in mouse models of helminth infection and colitis. J Exp Med 206:655–667
von Freeden-Jeffry U, Vieira P, Lucian LA, McNeil T, Burdach SEG, Murray R (1995) Lymphopenia in interleukin (IL)-7 gene-deleted mice identifies IL-7 as a nonredundant cytokine. J Exp Med 181:1519–1526
Vosshenrich CA, Cumano A, Muller W, Di Santo JP, Vieira P (2003) Thymic stromal-derived lymphopoietin distinguishes fetal from adult B cell development. Nat Immunol 4:773–779
Yoshida H, Honda K, Shinkura R, Adachi S, Nishikawa S, Maki K, Ikuta K, Nishikawa SI (1999) IL-7 receptor alpha+ CD3(−) cells in the embryonic intestine induces the organizing center of Peyer's patches. Int Immunol 11:643–655

# Part II
# Development of Lymph Nodes in Humans and Rodents

# Chapter 6
# Developmental Relationship and Convergence Between the Formation of Lymphoid Organs and Lymphatic Vasculature

**Péter Balogh**

**Abstract** The lymphatic system ensures a continuous flow of interstitial fluid and cell transfer through the lymph nodes and into the blood circulation. In addition, lymphatic capillaries also play an important role in the egress of lymphocytes from secondary lymphoid tissues. The specialization of lymphatic endothelium and its synchronization with the local blood vessel formation are crucial for proper lymph node circulation. The embryonic appearance of lymphatic vessels on one hand is the result of a separate endothelial commitment but may also be initiated by nonendothelial cells under pathological conditions. This chapter outlines those fate-determining events, transcription factors, and endothelial growth factors that are necessary for the establishment of lymphatic endothelium identity, and the relationship between the lymphatic vessel development and lymph node formation.

## 6.1 Introduction

Functions of the lymphatic system as a parallel circulatory network to the blood circulation include intestinal lipid absorption, interstitial protein balance and fluid homeostasis, and antigen transport to lymphoid organs. These vessels also have important roles in malignancies and inflammations. In addition to controlling fluid homeostasis, lymphatics provide opportunity for the immune system to sample the internal environment for pathogenic entry by the continuous transport of lymph through lymph nodes, which process constitutes an important part of immunological surveillance by the majority of lymphocytes residing in peripheral lymphoid tissues.

Lymphatic endothelial cells (LECs) lining the lymphatic vessels differentiate from venous endothelium (Sabin 1902); thus, they represent a specialized branch

P. Balogh
Department of Immunology and Biotechnology, Faculty of Medicine, University of Pécs, Pécs, Hungary

P. Balogh (ed.), *Developmental Biology of Peripheral Lymphoid Organs*,
DOI 10.1007/978-3-642-14429-5_6, 

of endothelial lineages. This divergence of peripheral endothelium recalls an intriguing parallel to that of the hemopoietic lineages within the bone marrow, indicating that the capacity of differentiating along various lineages is preserved within both the endothelial and hemopoietic descendants of common hemangioblasts (Kubo and Alitalo 2003). Interestingly, in addition to lymphatic vessels, lymph nodes and several other peripheral lymphoid organs also contain venules that are lined by specialized set of endothelium, whose expression is restricted to these tissues under normal circumstances. Recent observations that describe potential transdifferentiation of macrophages into LECs hint at an even closer relationship between the two branches (see below).

Historically, the development of lymphatic system was interpreted along two hypotheses. Proposed by Florence Sabin over a hundred years ago, the "centrifugal" way predicted that, after budding from the cardinal vein, a set of endothelial cells will give rise to the lymphatic vasculature. In contrary, the "centripetal" hypothesis of Huntington and McClure (Huntington and McClure 1910) implied the existence of lymphangioblasts which, in a scattered fashion, give rise to the regional lymphatics, which will subsequently join each other and to the venous circulation. It is still debatable that in various classes of vertebrates which mechanism is dominant, and whether there are alternative sources or processes for lymphatic vessel formation.

Another important developmental connection between the lymphatic vessels and peripheral lymphoid organs is the site of beginning of lymph node formation. Classic embryological observations revealed that the initiation of lymph node development is closely related to the formation of lymph sacs at the early stage of lymphatic vasculature development (for details see Chap. 7). Here the clustering of hemopoietic cells and expansion of undifferentiated mesenchyma also involves the transformation of local lymphatic endothelium to form the subcapsular sinus, an important platform implicated in sampling of lymph-borne antigens postnatally.

The presence of differentiated LECs, however, does not necessarily accompany with the subsequent development of lymph nodes, as evidenced in vertebrates that possess lymphatic vasculature, but no lymph nodes; on the other hand, it is not yet known to what extent the lymph node formation depends on the previous differentiation of lymphatic vasculature.

Last, ongoing immune responses as well as pathological events (chronic inflammation or malignant tumors) influence the structure of already existing lymphatic vessels and may initiate the formation of new ones, which may either prolong the persistence of the process, or facilitate its spreading to other regions. Other pathological conditions of LECs coupled with lymphoid organ development and immunological competence include Kaposi sarcoma, possibly representing malignant transformation of (preferentially) LECs or their committed/inducible precursors upon infection with human herpesvirus-8 (HHV-8), a condition most often associated with acquired immunodeficiency syndrome (AIDS; Jussila et al. 1998; Hong et al. 2004).

The present chapter overviews the structural and developmental features of lymphatic vessels and their relationship with the formation of lymph nodes.

## 6.2 Phylogeny of Lymphatic Vessels

Traditionally, the features of the vertebrate lymphatic vasculature have been studied most extensively in mammals. Vessels with structural and functional similarities to mammalian lymphatics can also be found in lower vertebrates, such as amphibians, where the parallel lymphatic circulation also contains a lymph heart. The increase of animal's size may have necessitated the separation of the blood circulation from the lymphatic circulation, making former the dominant form of supplying nutrients and gas exchange, whereas the role of lymphatic circulation became restricted to interstitial fluid homeostasis, leukocyte recirculation, and antigen dispersal (Jeltsch et al. 2003). In lizards (reptiles), the capacity to generate new lymphatic vessels after birth significantly contributes to the tail regeneration following its autotomy as part of predator escape reaction (Daniels et al. 2003). Lymphatic vessels are also present in birds, although largely with no lymph nodes developed yet. This discrepancy indicates that the formation of lymph nodes as filter elements of the lymphatic circulation is a relatively recent phenomenon in phylogeny. In addition, in some mutant mice where the development of lymphatic vasculature is blocked, embryonic lymph node anlage may form (Vondenhoff et al. 2009). Interestingly, the earliest vessels in vertebrate embryos also express markers and transcription factors characteristic for LECs (see below).

## 6.3 Structure of Lymphatic Vessels in Mammals

The structure of lymphatic vessels was shaped by its preimmune functions, mostly the rapid exchange and channeling of interstitial liquid (from capillary extravasation) towards the venous system in a one-way direction which connects to the venous circulation. Accordingly, the lymphatic capillaries are lined by a thin layer of LECs, with considerably looser intercellular connections compared with blood endothelial cells (BECs) both to each other and their surroundings. Instead of even surface and possessing tight lateral connections (tight junctions and adherent junctions) as in BECs, LECs show roof-tile arrangement, with the overlapping segments forming the primary valves that facilitate the one-way transport of lymph (Jeltsch et al. 2003). Larger caliber lymphatic vessels have valves. In addition to looser interendothelial connections, initial lymphatics have no continuous basement membrane and surrounding smooth muscle cells. Instead of basement membrane/cellular anchorage, LECs are tethered to the extracellular matrix via anchoring filaments. These structures contain fibrillin with the RGD motif to be recognized by $\alpha v\beta 3$ integrin expressed in LECs. Depending on the pressure of interstitial fluid, these anchoring filaments transmit the mechanical drive for open or close positioning of LECs, thus regulate the fluid uptake into lymphatic capillaries. Larger collecting lymphatics have basement membrane and contractile pericyte cover which, together with the skeletal muscle activity and arterial pulsation, assist in the propulsion of lymph to empty into the venous circulation via the *ductus thoracicus* and the *ductus lymphaticus dexter* at the junction between the subclavian vein and the jugular vein on the left and right, respectively.

## 6.4 Development of Lymphatic Vessels: Programmed or Induced

The development of lymphatic vasculature represents a step-by-step gradual commitment and specification towards *one* possible progeny of endothelial lineages; thus, the general vascular differentiation of the organism to a large extent overlaps with the establishment of LEC fate and lymphatic vessel characteristics. This process includes the emergence of endothelium, commitment along the venous specification, as well as the induction and maintenance of lymphatic endothelium identity. Parallel to the growth of embryo, the lymphatic vasculature also needs to expand and undergo vessel remodeling in relationship with its tissue microenvironment. Recent progress in screening for genes with differential expression (with approximately only 2% difference) between vascular endothelium and LECs, together with developing reagents that identify markers restricted to LECs and creating mice with targeted mutagenesis of genes implicated in LEC development, has greatly expanded our understanding the details of these processes (Hirakawa et al. 2003).

## 6.5 Embryonic Development: Establishment of LEC Identity

As part of the general vascular development, the formation of lymphatics is initiated by the commitment towards endothelium to commence vasculogenesis, which will give rise to those cardinal veins that serve as subsequent starter sites for LEC specification. This entire process can be divided into the following stages (1) differentiation of endothelium, (2) vasculogenesis, (3) establishment of arterial or venous endothelial identity, (4) induction of LEC differentiation/competence in cardinal vein vascular endothelial cells and their budding, and finally (5) formation of lymphatic vessels from lymphatic sacs.

The initiation of endothelial differentiation is coupled to sequential differentiation of mesenchyme into perivascular smooth cells/pericytes and into common hemogenic endothelium, giving rise to blood cells and endothelium. The putative common precursors probably express VEGFR-2 and are the subjects of the combined effect of VEGF binding and basic helix-loop-helix (bHLH) transcription factors Tal-1/Scl. Depending on the expression level of Tal-1, VEGFR-$2^+$/Tal-$1^{hi}$ cells are committed along endothelial specification, whereas VEGFR-$2^+$/Tal-$1^{lo}$ cells will give rise to pericytes (Ema et al. 2003; D'Souza et al. 2005) in a process which also involves Wnt pathway (Shin et al. 2009). Proper differentiation and clustering of pericytes with developing endothelial cells involves the signaling via platelet-derived growth factor (PDGF-R) upon binding PDGF-β, whereas VEGFR activation negatively regulates PDGF-R-mediated signaling, thus blocks pericyte differentiation (Greenberg et al. 2008). The full differentiation capacity of hemangioblasts and their eventual developmental paths in connection with hemopoietic cell formation are, however, controversial and largely unexplored yet. These cells may first develop into the intermediary form of "hemogenic endothelium," which will later develop along the endothelial or hemopoietic cells in a process involving Runx1 transcription factor subsequent to Tal-1/Scl expression, in addition to forming other vessel-associated elements as well (Lancrin et al. 2009).

The differentiation of endothelial cells is followed by their expansion and arrangement into complex vascular network. These latter processes are promoted by the ligand binding of VEGF receptors (VEGFR-1/flt-1, VEGFR-2/flk-1, and VEGFR-3/flt-4), with a perplexing array of recognition pattern. Thus VEGF-R1 binds VEGF-A and VEGF-B, VEGFR-2 binds VEGF-A/C/D, and VEGFR-3 binds VEGF-C/D. For endothelial cell survival, the most important interaction may be conferred by the VEGF-A/VEGFR-2 signaling, as mice mutant for these components died earliest, followed by mice rendered deficient for VEGF-R3, whereas the lethal effect of the targeted mutation of VEGFR-1 manifested latest (Hosking and Makinen 2007). This last mutation, however, severely impaired the commitment towards LEC development. In addition to VEGF binding by VEGFRs, these cytokines may also be recognized by neurolipin-1 and neurolipin-2, two nonkinase type receptors (Nrps). Importantly, for LEC differentiation, Nrp-2 is primarily expressed in veins and lymphatics and may bind VEGF-C (similarly to VEGFR-3), whereas arterial endothelial cells produce Nrp-1 (Oliver and Srinivasan 2008; Jurisic and Detmar 2009).

The vascular formation at a later stage is also assisted by the angiopoietin family members (Ang 1–4) produced dominantly by smooth muscle cells, and their endothelium-specific receptor Tie-2/Tek, primarily by Ang-1-Tie-2 engagement, which promotes proper arrangement between the endothelium and its mural neighboring cells. The relationship between Ang1/Ang2 binding to Tie-2 is yet to be determined; it is probable that while they act in an antagonistic manner in blood vessels, they function as agonists in lymphatic vessel formation Oliver and Srinivasan 2008).

Subsequent to the emergence/expansion of endothelial cells and vascular stabilization, next their arterial or venous identity will be defined by at least three sets of morphogenic regulators. These include Notch receptor family (Notch 1–4) and their ligands, Jagged 1 & 2 and Delta 1, 3, & 4. Expression of Notch 1 and 4 on arterial endothelium and their ligands, and Notch-2 on smooth muscle cells surrounding these cells, and their absence on venous segments, respectively, indicate that these factors promote arterial commitment and also arterial vessel structural integrity. Another group of regulators include ephrins and their Eph receptor family, comprising at least 15 members with tyrosine kinase activity. Depending on the type of vasculature, Eph-B2-type receptor in mice selectively identifies arterial endothelium, whereas Eph-B4 is expressed by venous endothelium (Adams et al. 1999). Finally, venous commitment is also promoted by the expression of chicken ovalbumin upstream promoter transcription factor II (COUP-TFII), whose absence redirects the venous endothelium to gain characteristics of arterial endothelium (You et al. 2005). In addition to these subset-specific factors, other regulatory circuits include genes associated with morphogenesis, such as the members of the HOX gene cluster (HOX A, HOX B, HOX C, and HOX D), with effects ranging from cell-autonomous endothelial commitment through multicellular vasculogenesis (Pruett et al. 2008).

With the formation of anterior cardinal vein, the stage for the induction of LEC commitment is set. A primary factor for commitment is the Prox1 (Prospero-related homeodomain factor-1) transcription factor, which exerts its effects in competent venous endothelial cells, marked by their expression of LYVE-1 hyaluronan receptor (Wigle and Oliver 1999; Wigle et al. 2002). Prox1 induction at one segment of the cardinal veins indicates the endothelial cells' commitment along LEC specification. Inactivation of Prox1 blocks LEC differentiation, whereas its enhanced expression causes vascular endothelial cell reversal towards LEC lineage, as evidenced by the expression of mature LEC-specific markers (Hong et al. 2002). Although it constitutes a key event during the development of LECs, it is not yet known what stimuli are involved in the upregulation of Prox1-expression. A potential direct downstream target gene for Prox1 is the Type 3 receptor for fibroblast growth factor (FGF-R3), binding FGF-1 & 2 (Shin et al. 2006). Endothelial cells with the VEGFR-3$^{+}$/LYVE-1$^{+}$/Prox1$^{+}$ phenotype will bud from the embryonic veins and, probably upon the effect of VEGF-C binding by VEGFR-3, they will migrate away from the veins to form lymph sacs (Oliver and Srinivasan 2008). This process therefore involves (a) a cellular programming towards LEC identity and (b) the physical separation of LEC-primed cells from the cardinal veins, and their repositioning to distal regions. This latter event is referred to as lymphovenous separation. The expression of VEGFR-3 by embryonic venous endothelium is downregulated in later embryos, while in LECs, it is preserved even in the postnatal period (Oliver and Srinivasan 2008).

Subsequent to the segregation of lymphatic sacs, further expansion/proliferation and remodeling of LECs are required for the formation of lymphatic network which, at certain predefined locations, also involves the formation of lymph nodes. However, only rudimentary information is available as yet on the endothelial regulators involved in these events. VEGFR-3 binding of VEGF-C constitutes a major regulatory pathway, resulting in an extensive sprouting of lymphatic capillaries. Similarly to the blood vessels, the lymphatic capillaries are stabilized by the participation of angiopoietins 1 & 2 (Jurisic and Detmar 2009). This mechanism for the stabilization is required for the proper "transversal" organization of lymphatic structure, whereas in the "longitudinal" (capillary or collecting-type lymphatics) specification, FOXC2 forkhead transcription factor plays an important role, with its absence resulting in the lack of valves (Petrova et al. 2004). Once the developmental programming has been completed, both in adult humans and mice lymphatic capillaries can be identified by their expression of LYVE-1, podoplanin (a

**Table 6.1** Main events in the development of lymphatics

| Event | Target cell | Mediator/action | Effect |
|---|---|---|---|
| Formation of angioblasts | Hemangioblast or hemogenic endothelium | Scl/Tal1 and Runx1 transcription factors | Commitment along endothelial specification |
| Vasculogenesis | Undifferentiated endothelium | VEGF binding to VEGFRs and Nrp's; Signaling via Tie-2 by Angiopoietins 1–4 | Formation and stabilization of larger blood vessels |
| Arterial and venous segregation | Endothelium, smooth muscle cells and pericytes | Notch receptors and their ligands (Jagged, Delta); ephrin binding to Eph-B2 (artery); COUP-TFII and ephrin binding by Eph-B4 (vein). | Determination of arterial and venous identity of developing vessels |
| Induction of commitment along LEC lineage | Endothelial cells of cardinal veins expressing LYVE-1 and VEGFR-3 | Prox1 transcription factor | Induction/stabilization of LEC-associated gene expression |
| Formation of lymphoid sacs | LEC precursors | VEGFC binding to VEGFR-3 | Shedding from the cardinal veins |
| Sprouting of lymphoid capillaries | LECs in lymphoid sacs | VEGFR-3 and VEGF-C; angiopoietins via Tie-2; FOXC2 transcription factor | Expansion of LECs from lymphatic vessels; stabilization of structure and separation of lymphatic capillaries and larger collecting vessels |

transmembrane glycoprotein also shared with T-zone reticular fibroblasts in peripheral lymphoid tissues) and VEGFR-3 surface markers, and the production of Prox1 transcription factor. The main events during embryonic LEC commitment and differentiation together with their presumed functions are summarized in Table 6.1.

## 6.6 Inflammatory Macrophages: Assist or Impersonate

In various pathological conditions (inflammation or malignancy), de novo differentiation of lymphatic vasculature can be observed. These vessels may form as extensions of preexisting lymphatic capillaries, but may also be generated from other cells. As the subject of this volume is primarily dedicated to the developmental processes of peripheral lymphoid tissues, the readers interested in malignancy-related lymphangiogenesis are referred to other sources.

Inflammation represents both physiological defense reaction against invading pathogens in infections or may be associated with tissue-specific autoimmune diseases or tumor progression. In the process of de novo formation of lymphatic vessels, the spotlight has focused mainly on two cell types. Of these, activated B cells may secrete large amount of VEGF-A, a potent growth factor for lymphatic vessel sprouting in lymph nodes (Angeli et al. 2006). In lymph nodes, the lymphatic vessels and HEVs jointly undergo remodeling processes during immune responses in a B cell-dependent and lymphotoxin β-receptor (LTβR)-mediated manner (Liao and Ruddle 2006).

The other hemopoietic cells that have been closely associated with LEC differentiation are the macrophages (Xiong et al. 1998). Moreover, unlike B cells, these latter cells are in a dual relationship with LEC cells, as they can both produce VEGF-C, an important LEC-selective ligand for VEGFR-3, and become LECs themselves upon transdifferentiation by expressing VEGFR-3 (Maruyama et al. 2005). During expansion, LECs may further increase their CCL21 chemokine production (Martin-Fontecha et al. 2003); thus, the enhanced recruitment of more dendritic cells and lymphocytes may establish a positive loop for (tertiary) lymphoid tissue stabilization and enlargement (Johnson and Jackson 2008). In addition, other cells (including nonmyeloid committed hemopoetic stem cells) may also give rise to LECs under noninflammatory conditions (Jiang et al. 2008). Future studies will determine whether the macrophage-derived LEC formation described for murine corneal lymphatic vessel development may also be applicable for human secondary lymphoid tissues.

## References

Adams RH, Wilkinson GA, Weiss C, Diella F, Gale NW, Deutsch U, Risau W, Klein R. (1999) Roles of ephrinB ligands and EphB receptors in cardiovascular development: demarcation of arterial/venous domains, vascular morphogenesis, and sprouting angiogenesis. Genes Dev 13:295–306

Angeli V, Ginhoux F, Llodrà J, Quemeneur L, Frenette PS, Skobe M, Jessberger R, Merad M, Randolph GJ (2006) B cell-driven lymphangiogenesis in inflamed lymph nodes enhances dendritic cell mobilization. Immunity 24:203–215

Daniels CB, Lewis BC, Tsopelas C, Munns SL, Orgeig S, Baldwin ME, Stacker SA, Achen MG, Chatterton BE, Cooter RD (2003) Regenerating lizard tails: a new model for investigating lymphangiogenesis. FASEB J 17:479–481

D'Souza SL, Elefanty AG, Keller G (2005) SCL/Tal-1 is essential for hematopoietic commitment of the hemangioblast but not for its development. Blood 105:3862–3870

Ema M, Faloon P, Zhang WJ, Hirashima M, Reid T, Stanford WL, Orkin S, Choi K, Rossant J (2003) Combinatorial effects of Flk1 and Tal1 on vascular and hematopoietic development in the mouse. Genes Dev 17:380–393

Greenberg JI, Shields DJ, Barillas SG, Acevedo LM, Murphy E, Huang J, Scheppke L, Stockmann C, Johnson RS, Angle N, Cheresh DA (2008) A role for VEGF as a negative regulator of pericyte function and vessel maturation. Nature 456:809–813

Hirakawa S, Hong YK, Harvey N, Schacht V, Matsuda K, Libermann T, Detmar M (2003) Identification of vascular lineage-specific genes by transcriptional profiling of isolated blood vascular and lymphatic endothelial cells. Am J Pathol 162:575–586

Hosking B, Makinen T. (2007) Lymphatic vasculature: a molecular perspective. Bioessays 29:1192–1202
Hong YK, Harvey N, Noh YH, Schacht V, Hirakawa S, Detmar M, Oliver G (2002) Prox1 is a master control gene in the program specifying lymphatic endothelial cell fate. Dev Dyn 225:351–357
Hong YK, Foreman K, Shin JW, Hirakawa S, Curry CL, Sage DR, Libermann T, Dezube BJ, Fingeroth JD, Detmar M (2004) Lymphatic reprogramming of blood vascular endothelium by Kaposi sarcoma-associated herpesvirus. Nat Genet 36:683–685
Huntington GS, McClure CFW (1910) The anatomy and development of the jugular lymph sac in the domestic cat (Felis domestica). Am J Anat 10:177–311
Jeltsch M, Tammela T, Alitalo K, Wilting J (2003) Genesis and pathogenesis of lymphatic vessels. Cell Tissue Res 314:69–84
Jiang S, Bailey AS, Goldman DC, Swain JR, Wong MH, Streeter PR, Fleming WH (2008) Hematopoietic stem cells contribute to lymphatic endothelium. PLoS One 3:e3812
Johnson LA, Jackson DG (2008) Cell traffic and the lymphatic endothelium. Ann N Y Acad Sci 1131:119–133
Jurisic G, Detmar M (2009) Lymphatic endothelium in health and disease. Cell Tissue Res 335:97–108
Jussila L, Valtola R, Partanen TA, Salven P, Heikkila P, Matikainen MT, Renkonen R, Kaipainen A, Detmar M, Tschachler E, Alitalo R, Alitalo K (1998) Lymphatic endothelium and Kaposi's sarcoma spindle cells detected by antibodies against the vascular endothelial growth factor receptor-3, Cancer Res 58:1599–1604
Kubo H, Alitalo K (2003) The bloody fate of endothelial stem cells. Genes Dev 17:322–329
Lancrin C, Sroczynska P, Stephenson C, Allen T, Kouskoff V, Lacaud G (2009) The haemangioblast generates haematopoietic cells through a haemogenic endothelium stage. Nature 457:892–895
Liao S, Ruddle NH (2006) Synchrony of high endothelial venules and lymphatic vessels revealed by immunization. J Immunol 177:3369–3379
Martin-Fontecha A, Sebastiani S, Höpken UE, Uguccioni M, Lipp M, Lanzavecchia A, Sallusto F (2003) Regulation of dendritic cell migration to the draining lymph node: impact on T lymphocyte traffic and priming. J Exp Med 198:615–621
Maruyama K, Ii M, Cursiefen C, Jackson DG, Keino H, Tomita M, Van Rooijen N, Takenaka H, D'Amore PA, Stein-Streilein J, Losordo DW, Streilein JW (2005) Inflammation-induced lymphangiogenesis in the cornea arises from CD11b-positive macrophages. J Clin Invest 115:2363–2372
Oliver G, Srinivasan RS (2008) Lymphatic vasculature development: current concepts. Ann N Y Acad Sci 1131:75–81
Petrova TV, Karpanen T, Norrmén C, Mellor R, Tamakoshi T, Finegold D, Ferrell R, Kerjaschki D, Mortimer P, Ylä-Herttuala S, Miura N, Alitalo K (2004) Defective valves and abnormal mural cell recruitment underlie lymphatic vascular failure in lymphedema distichiasis. Nat Med 10:974–981
Pruett ND, Visconti RP, Jacobs DF, Scholz D, McQuinn T, Sundberg JP, Awgulewitsch A (2008) Evidence for Hox-specified positional identities in adult vasculature. BMC Dev Biol 8:93
Sabin FR (1902) On the origin of the lymphatic system from the veins and the development of the lymph hearts and thoracic duct in the pig. Am J Anat 1:367–391
Shin JW, Min M, Larrieu-Lahargue F, Canron X, Kunstfeld R, Nguyen L, Henderson JE, Bikfalvi A, Detmar M, Hong YK (2006) Prox1 promotes lineage-specific expression of fibroblast growth factor (FGF) receptor-3 in lymphatic endothelium: a role for FGF signaling in lymphangiogenesis. Mol Biol Cell 17:576–584
Shin M, Nagai H, Sheng G (2009) Notch mediates Wnt and BMP signals in the early separation of smooth muscle progenitors and blood/endothelial common progenitors. Development 136:595–603

Vondenhoff MF, van de Pavert SA, Dillard ME, Greuter M, Goverse G, Oliver G, Mebius RE (2009) Lymph sacs are not required for the initiation of lymph node formation. Development 136:29–34

Wigle JT, Oliver G (1999) Prox1 function is required for the development of the murine lymphatic system. Cell 98:769–778

Wigle JT, Harvey N, Detmar M, Lagutina I, Grosveld G, Gunn MD, Jackson DG, Oliver G (2002) An essential role for Prox1 in the induction of the lymphatic endothelial cell phenotype. EMBO J 21:1505–1513

Xiong M, Elson G, Legarda D, Leibovich SJ (1998) Production of vascular endothelial growth factor by murine macrophages: regulation by hypoxia, lactate, and the inducible nitric oxide synthase pathway. Am J Pathol 153:587–598

You LR, Lin FJ, Lee CT, DeMayo FJ, Tsai MJ, Tsai SY (2005) Suppression of Notch signalling by the COUP-TFII transcription factor regulates vein identity. Nature 435:98–104

# Chapter 7
# Development and Structure of Lymph Nodes in Humans and Mice

**Tom Cupedo, Mark C. Coles, and Henrique Veiga-Fernandes**

**Abstract** Throughout the human body, 500–600 lymph nodes are situated. These secondary lymphoid organs collect antigens from peripheral tissues via the afferent lymphatics and provide T and B lymphocytes with the optimal environment for cellular activation and proliferation. In this chapter, we will highlight the interactions between hematopoietic cells and stromal cells that are essential for the proper formation and organization of the lymph node. In addition, the distinct cellular architecture that is characteristic for secondary lymphoid organs as well as its vascular system will be discussed in the context of lymph node function.

## 7.1 Secondary Lymphoid Organs

The prime function of secondary lymphoid organs is to increase the efficiency of immune responses through the generation of high affinity antibodies and by serving as a platform for naive T cells to encounter their cognate antigen presented by professional antigen presenting cells (APC) (Crivellato et al. 2004). B cell activation and differentiation into plasma or memory cells takes place within the germinal center, a specialized microdomain within the B cell follicle in which B cells are allowed to mutate the variable region of their rearranged immunoglobulin gene with the ultimate goal of achieving high-affinity antibodies (Kelsoe 1996; Allen

---

T. Cupedo (✉)
Department of Hematology, Erasmus University Medical Center, Rotterdam, The Netherlands
e-mail: t.cupedo@erasmusmc.nl

M.C. Coles
Centre for Immunology and Infection, Department of Biology and Hull York Medical School, University of York, York, UK

H. Veiga-Fernandes
Immunobiology Unit, Instituto de Medicina Molecular, Faculdade de Medicina de Lisboa, Lisboa, Portugal

P. Balogh (ed.), *Developmental Biology of Peripheral Lymphoid Organs*,
DOI 10.1007/978-3-642-14429-5_7, 

et al. 2007; Klein et al. 2008). Since introducing germline mutations harbors the intrinsic risk of malignant transformations survival of B cells during this process is therefore strictly dependent on intimate interactions with follicular dendritic cells (FDCs), the most abundant stromal component of the germinal center (Klein et al. 2008; Schmidlin et al. 2009). FDCs are thought to develop from the local mesenchyme within the follicle, and their development is initiated by B cell-derived lymphotoxin (LT) and TNF (Endres et al. 1999; van Nierop et al. 2002). Sufficiently high levels of LT and TNF are only achieved at sites where there is an accumulation of large numbers of B cells: the primary B cell follicle. In addition, B cell activation critically depends on additional signals delivered by T cells. This dependency of B cells on lymphoid follicles and activated T cells implies that germinal center reactions and the generation of high affinity antibodies will only occur in structurally organized organs, highlighting the importance of the lymphoid organ architecture for immunological responses. In contrast, T cell dependence on secondary lymphoid organs seems to be less stringent, and T cell activation can occur away from secondary lymphoid organs in, for instance, the liver (Greter et al. 2009). However, by bringing together dendritic cells (DC) present antigens sampled from discrete areas of the body with recirculating naïve T cells, the chance that a T cell finds the very few DCs that carry its cognate antigen is dramatically increased. Lastly, within the secondary lymphoid organs activated T cells are also endowed with some level of regional specificity, as T cells activated in peripheral lymphoid organs will express homing molecules allowing them to preferentially return to the drainage area of that lymph node, where the infection is likely to be occurring (Dudda and Martin 2004; Mora and von Andrian 2006).

The highly organized structure of secondary lymphoid tissues is dictated by different stromal cell networks that facilitate the initiation and maintenance of antigen-specific immune responses. These mesenchymal networks include fibroblast reticular cells (FRCs) that support T cell migration and T cell–DC interactions, FDCs which present antigen to B cells, follicular stromal cells (FSCs) that form a network within B cell follicles and can give rise to FDCs and marginal reticular cells (MRCs), which are associated with subcapsular sinus (SCS) macrophages (Katakai et al. 2004; Katakai et al. 2004; Bajénoff et al. 2006; Link et al. 2007; Katakai et al. 2008; Mueller and Germain 2009). Although all of these different lymph node stromal populations are thought to be mesenchymal in origin, the lineage relationship between the different stromal populations is currently unclear. Development and organization of these stromal networks results from interactions between mesenchymal stromal cells and hematopoietic cells including B and T cells during the postnatal period and lymphoid tissue inducer (LTi) cells during fetal development (Katakai et al. 2004; White et al. 2007).

LTi cells are essential for the fetal induction of lymph nodes and Peyer's patches in both humans and mice (Mebius 2003). In the developing mouse embryo, LTi cells cluster at sites of prospective lymph node development, where they induce the local mesenchyme to differentiate into lymphoid tissue organizer (LTo) cells through LT-beta receptor and TNF receptor triggering (Yoshida et al. 1999;

White et al. 2007; van de Pavert et al. 2009; Vondenhoff et al. 2009). As a result of this interaction, stromal cells will produce chemokines such as CXCL13 and CCL21 and induce expression of adhesion molecules including ICAM-1, VCAM-1, and MAdCAM-1. This early lymph node primordium is now equipped to attract and retain circulating cells. These will include LTi cells and their putative precursors but also additional hematopoietic cells that are present in the embryo including B cells, NK cells, and myeloid cells (Mebius et al. 1996; Mebius et al. 1997). Influx of these cells will enforce LTβR and TNFR signaling leading to the formation of a self-sustaining primordium that will eventually develop into a lymph node (Mebius 2003).

In both humans and mice, lymph nodes always develop at fixed positions in the body. This indicates that positioning is likely to be determined by the induction of a specific gene program in nonmotile cells at predetermined sites. It has long been thought that the initiating signal for lymph node development is delivered by the lymphatic endothelium. During early lymph node development, the vascular endothelium of the veins starts to express the homeobox gene *Prox1* at strictly defined regions of the body (Wigle et al. 1999). These Prox-1 positive endothelial cells in subsequent developmental steps bud of from the vein to form a lymph sac. This lymph sac will eventually develop into the subcapsullar sinus of the lymph node and will give rise to the first lymphatic vessels through a sprouting process. However, careful analysis of the earliest lymph node primordia have made clear that this process of endothelial cell specification, while being essential for correct formation of the lymph node as well as the lymphatic vasculature, is not required for the initial aggregation of LTi cells and LTo cells (Vondenhoff et al. 2009). Recent data suggest that the first production of chemokines responsible for LTi cell attraction at sites of lymph node development is regulated by neuronal-derived signals (van de Pavert et al. 2009). Whether neuronal signals are involved in Peyer's patch genesis is currently unknown.

In this chapter, the formation of lymph nodes will be discussed, with a focus on the stromal and hematopoietic cells involved in this process, the organization of the lymphoid organs into the characteristic segregation of T and B cells and the functional consequences of this typical architecture.

## 7.2 Hematopoietic Cells in Lymph Node Development

LTi cells have been identified in both humans and mice (Kelly and Scollay 1992; Mebius et al. 1997; Cupedo et al. 2009). LTi cells are identified by expression of RORγt in conjunction with the IL7R alpha chain (CD127), surface Lymphotoxinα1β2 (LT), and the absence of lineage markers (Mebius et al. 1997; Sun et al. 2000; Eberl et al. 2004; Cupedo et al. 2009). The finding that mice with a targeted disruption of the LTβR signaling pathway fail to develop lymph nodes and Peyer's patches signified the absolute necessity of this pathway for correct lymphoid organ development (De Togni et al. 1994; Banks et al. 1995; Rennert et al. 1996; Alimzhanov et al. 1997). The regulation of surface LT expression on fetal LTi cells is therefore of crucial importance and was shown to be controlled by two partially overlapping signals. IL-7-receptor ligation is needed for LT induction on LTi cells in the Peyer's patches but is dispensable for lymph node development. On the other hand, RANK activation by RANKL is needed for the development of most lymph nodes, while this signal is not needed for Peyer's patch development (Yoshida et al. 2002).

LTi cells remain present in adults and have been most extensively studied in adult murine spleen and intestines (Kim et al. 2003; Scandella et al. 2008; Tsuji et al. 2008). On the one hand, the function of adult LTi cells is related to organ remodeling and as such is analogous to the function of their fetal counterparts. In the intestines, LTi cells reside mainly in crypopatches and are needed for the transformation of cryptopatches into isolated lymphoid follicles, intestinal sites of IgA production that form postnatal (Tsuji et al. 2008). In the postnatal spleen, LTi cells accelerate the reconstruction of the splenic cellular architecture after the clearance of destructive viral infections (Scandella et al. 2008). On the other hand, adult LTi cells play an active role in immunity and are important for the generation of T cell memory in the spleen by providing survival signals to memory T cells (Kim et al. 2003).

Development of murine LTi cells depends on the nuclear hormone receptor ROR$\gamma$t and the inhibition of E2A proteins by Id2 (Sun et al. 2000; Fukuyama et al. 2002; Eberl et al. 2004; Boos et al. 2007). LTi cell precursors reside in the fetal liver and are the fetal equivalent of the common lymphoid precursor that is present in adult mouse bone marrow (Mebius et al. 2001). The adult murine bone marrow contains cells capable of differentiating into LTi cells (Schmutz et al. 2009), although whether LTi cells are continuously replenished from the bone marrow has yet to be proven formally.

The first clues into signaling pathways involved in human lymph node development came from studies in which a limited number of patients with Severe Combined Immunodeficiency Disease (SCID) were assessed for the presence of lymph nodes post mortem. It became apparent that patients with mutations in the common gamma chain of cytokine receptors ($\gamma$c) or its downstream signaling kinase JAK3 lacked detectable lymph nodes (Facchetti et al. 1998). The facts that both $\gamma$c and JAK3 are essential components of the IL-7R signaling pathway, and that mice with deficiencies in this pathway lack lymph nodes, all support a role for IL-7R signaling in human lymph node development in analogy to the mouse (Park et al. 1995; Adachi et al. 1998). The likely importance of IL-7R signaling is underscored by high-level expression of the IL7R on human LTi cells. Human LTi cells were identified based on their phenotypic resemblance to murine LTi cells, their presence in fetal lymph nodes and their functional ability to interact with mesenchymal cells in vitro (Cupedo et al. 2009).

## 7.3 Stromal Cells in Lymph Node Development

During lymph node development in both humans and mice, mesenchymal cells at future lymph node locations differentiate into stromal LTo cells expressing VCAM-1, ICAM-1, and lymphotoxin Receptor β LTβR (Honda et al. 2001). The appearance of committed stromal LTo cells marks an essential event in the progression of lymph node formation determining the sites where lymph node organogenesis can occur. While development to this stage is not regulated by direct interactions with hematopoietic cells, the differentiation of localized mesenchyme into LNo cells requires paracrine stimulation. Utilizing mesenchymal stem cells (MSC) lines, which resemble tissue resident mesenchymal cells, it has been shown that incubation with recombinant TNFα or LTα1β2 was sufficient to lead to the upregulation of ICAM-1 and VCAM-1 on the MSCs (Ame-Thomas et al. 2007). Analysis of $LT\alpha^{-/-}$ lymph node anlagen from e16.5 embryos showed an essential role for LTα in the initial upregulation of VCAM-1, ICAM-1, and MAdCAM-1 on local mesenchymal cells, consistent with paracrine lymphotoxin signaling being important in LTo cell development (Vondenhoff et al. 2009). Although the source of TNF or LTα1β2 is unknown in developing embryos, one likely source is LECs that form the lymph sac abutting the developing anlagen. Although the lymph sac is not required for the initiation of lymph node development, conditional deletion of *Prox1* indicated that LECs have an important role in the maintenance of VCAM-1, MAdCAM-$1^{hi}$ cells (Vondenhoff et al. 2009).

In lymph nodes, sustained interactions between LTo cells and inducer cells meditated by signaling through the LTβR leads to the differentiation of LTo cells and induction of homeostatic chemokines (CCL19/21, CXCL13) and cytokines (IL-7) that attract and maintain the survival of LTi cells, lymphocytes, and DCs, VEGF-C which mediates the survival and growth of the lymphatic sinus and upregulation of RANK and LTβR expression required for the complete differentiation of LTo cells (Honda et al. 2001; Vondenhoff et al. 2009). In the absence of LTβR signaling, the initial formation of the anlagen occurs normally, yet the LTi cells fail to survive or move away from the developing anlagen and development of the lymph nodes is blocked at e16.5 in development (Coles et al. 2006). After birth, IL-7/IL7R interactions have an important role in mediating the maintenance of the maturing anlagen (Coles et al. 2006). Thus continual interactions between hematopoietic cells drive both the development and differentiation of lymph node stromal cells (Fig. 7.1).

Further differentiation in the postnatal period leads to the generation of T and B cells specific areas, defined by the presence of FRCs in the T cell areas, FSCs and FDCs in the B cell areas, and the appropriate patterns of specific chemokines and extracellular matrix (Mueller and Germain 2009). When lymph nodes are subsequently flooded with T and B cells, the characteristic architecture with strictly separated T and B cell areas is formed (Cupedo et al. 2004). B cells drive the further differentiation of FDCs from FSCs in the B cell follicles in a LT/TNF-dependent process, and their sustained chemokine production ensures the formation of defined

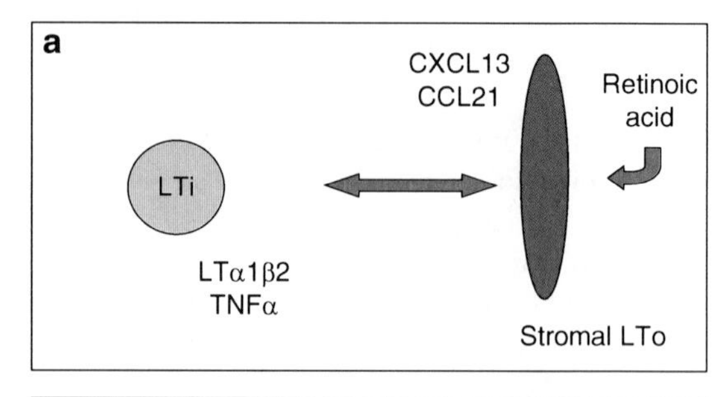


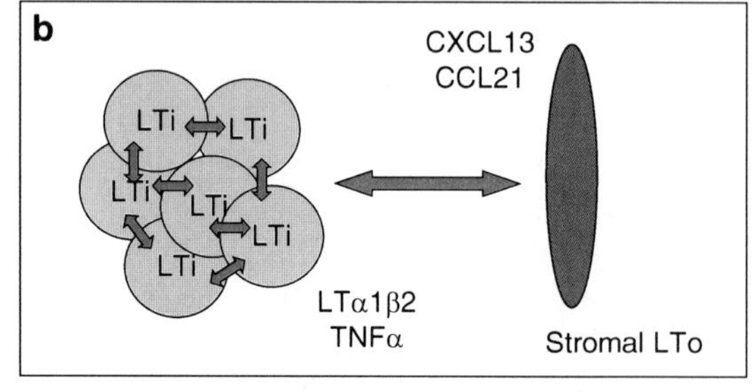


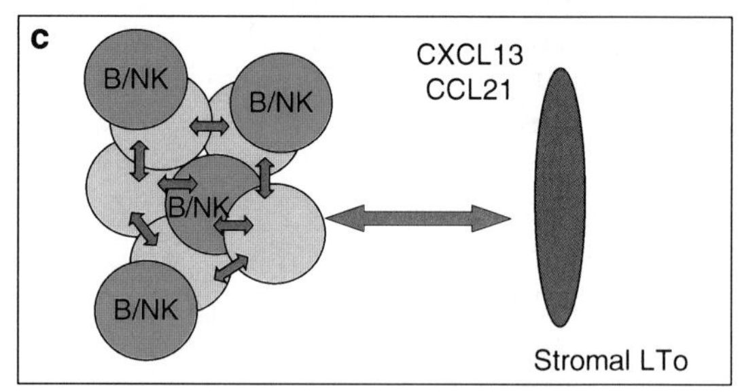


**Fig. 7.1** Cellular interactions during lymph node development. (**a**) Local production of RA by neurons induces stromal cells at future lymph node locations to secrete CXCL13 and CCL21. These chemokines attract LTi cells that will activate stromal cells via LTα1β2–LTβR and TNFα–TNFR interactions. (**b**) Continued production of CXCL13 and CCL21 will attract additional circulating LTi cells and their presumptive precursors. These will engage in paracrine interactions as well as in interactions with the stroma. (**c**) In the early postnatal stages, T and B cells will flood the lymph node and start to contribute to the stromal crosstalk to perfect the lymph node architecture

B cell follicles (Endres et al. 1999; Ansel et al. 2000). The continual relationship between hematopoietic cells and stromal cells is required for the establishment, organization, and function of stromal networks in lymph nodes.

## 7.4 Cellular Architecture in the Lymph Node

Probably, the single most characteristic feature of secondary lymphoid organs is their strict cellular architecture, in which T and B cells reside in anatomically distinct areas. In lymph nodes, B cells are located in the outer cortex, in spherical B cell follicles. T cells occupy the interfollicular areas as well as the deeper paracortical regions (Crivellato et al. 2004). In mice, this T/B cell segregation is a process that occurs after birth, when the first mature T cells leave the thymus and start to flood the lymph nodes (Cupedo et al. 2004). However, for both the lymph nodes and the Peyer's patches, rearranging lymphocytes are dispensable for the stromal specification of the T and B cell areas. In lymph nodes, all the lymphoid chemokines are induced normally in either SCID or RAG-deficient mice (Cupedo et al. 2004). Moreover, upon reconstitution of these immune-deficient mice with wild-type splenocytes, the transferred cells home to and lodge into their respective niches in the lymphoid organs (Gonzalez et al. 1998). Similarly, the Peyer's patches also form T and B cell areas based on stromal cell presence in the absence of lymphocytes (Hashi et al. 2001). CXCL13 (B lymphocyte chemo attractant) and CCL19 (EB/1 ligand chemokine/macrophage inflammatory protein-2-beta)/CCL21 (secondary lymphoid tissue chemokine) are opposing chemokines produced by stroma in B cell zone and T cell zones, respectively, controlling and guiding T and B cell positioning with in the lymphoid organs.

In adult mice, B cells express high levels of CXCR5 and are thus highly responsive to CXCL13 gradients, rapidly migrating from the high endothelial vessels (HEVs) in the T cell zone to the B cell zone. Resting B cells do not express CCR7 and are thus unresponsive to CCL19/21 expression in the T cell zone; however, antigen engagement of B cells leads to a rapid increase in CCR7 expression, leading to redistribution to the B–T cell interface (Reif et al. 2002). Stimulation of CXCR5 directly antagonizes CCR7 signaling; thus, the relative expression levels of both chemokines and chemokine receptors tightly define the localization of B cells within the adult lymph node. CXCL13 is produced by stromal cells in the B cells zones including FDCs in the B cell follicles and MRCs adjacent to the B cell zones.

In contrast, CCL19/21 is produced by FRCs and CCL21 by HEVs (Luther et al. 2000). The expression of CCL19/21 is required for the recruitment of T cells and DCs from both the afferent lymphatics through the HEVs. Control of CCR7 expression levels determines the retention of lymphocytes within the lymph node microenvironment.

Expression of both stromal-derived chemokines, CXCL13 and CCL21, can be downregulated during pathogen infections. Loss of CCL21 and CXCL13 expression was observed during viral and bacterial infections in draining lymph nodes in a $\gamma$-interferon-dependent process (Mueller et al. 2007). This downregulation may act to restrict the capacity of lymph nodes to respond to multiple different antigenic challenges; however, whether this is physiologically relevant is unclear as multiple mechanisms act to modulate rates of lymphocyte and antigen entry during immune responses.

The lymphoid tissue architecture of the lymph node becomes apparent postnatal upon influx of T cells, when secondary lymphoid organs go through several discrete stages of organization (Cupedo et al. 2004). Between days 2 and 4, B cells that were present scattered throughout the lymph node, relocate to the outer cortex, surrounding a central T cell zone. This T/B segregation is independent of the important B cell-chemokine CXCL13 as in mice deficient for this signaling pathway B cells arrange in the outer cortex normally. Only when both CXCR5 and CCR7 signaling are disturbed is the segregation of the lymphocyte subsets disrupted (Luther et al. 2003; Ohl et al. 2003). One day prior to the migration of B cells to the outer cortex, LTi cells are found to relocate to this area. While the signals regulating LTi cell migration or the functional significance of this migration are currently unknown, this distinctive migration pattern has led to some speculation on a possible role for LTi cells in inducing T/B cell segregation in the lymph nodes.

After day 4, B cells start to cluster, the earliest sign of follicle formation (Cupedo et al. 2004). The generation of lymphoid follicles is critically dependent on ligation of B cell-expressed CXCR5 by stromal cell-derived CXCL13. In the absence of CXCR5 signaling B cell follicles do not form and development is halted after T/B segregation (Ansel et al. 2000). Approximately, 1 week after birth FDCs develop within the follicles. FDCs are the main stromal subset within the follicle and are believed to be derived from the local mesenchyme. The appearance of FDCs marks the functional completion of the primary follicle that is now fully equipped to support germinal center reactions.

In contrast to mice, humans are born with fully developed lymph nodes. This difference in timing is a direct result from the fact that due to the longer pregnancy period in humans, T cell development is already completed in utero and the first mature T cells leave the thymus around week 12 of gestation (Pahal et al. 2000). Similar to murine development, B cells rearrange in the outer cortex by the 15th week of gestation followed by the appearance of FDC-containing B cell follicles between 17 and 19 weeks gestation (Westerga and Timens 1989; Westerga et al. 1989; Kasajima-Akatsuka and Maeda 2006; Cupedo et al. 2009).

## 7.5 Vasculature of the Lymph Nodes

Lymph nodes have two separate vascular systems: the lymphatic vasculature, which delivers antigen and APC from the peripheral tissues and organs to the draining lymph node via the afferent lymphatics and returns cells to the circulation via the efferent lymphatics; and the blood vasculature, which brings blood circulating lymphocytes, oxygen, and nutrients to the lymph node (Fu and Chaplin 1999). The lymph node has an extensive vascular network that enters from the medullary region and extends out to the paracortex. The lymph node is a high energy environment with continual lymphocyte migration, thus a requirement for high oxygen delivery. Multiphoton imaging has shown that in the absence of sufficient oxygen ($pO_2$) tension within the lymph node T cell motility and function is inhibited in an extremely rapid fashion (Huang et al. 2007). Thus, the two vascular systems, the lymphatic and blood vascular, are integral to understanding the function of the lymph node microenvironment (Fig. 7.2).

Tissue-resident T cells and DCs enter the lymph node through transmigration across the afferent lymphatics in a process dependent on adhesion through LFA-1, chemokine-mediated migration (CCL19/21), and S1P1-mediated signaling (Ledgerwood et al. 2008). This unidirectional migration appears mechanistically

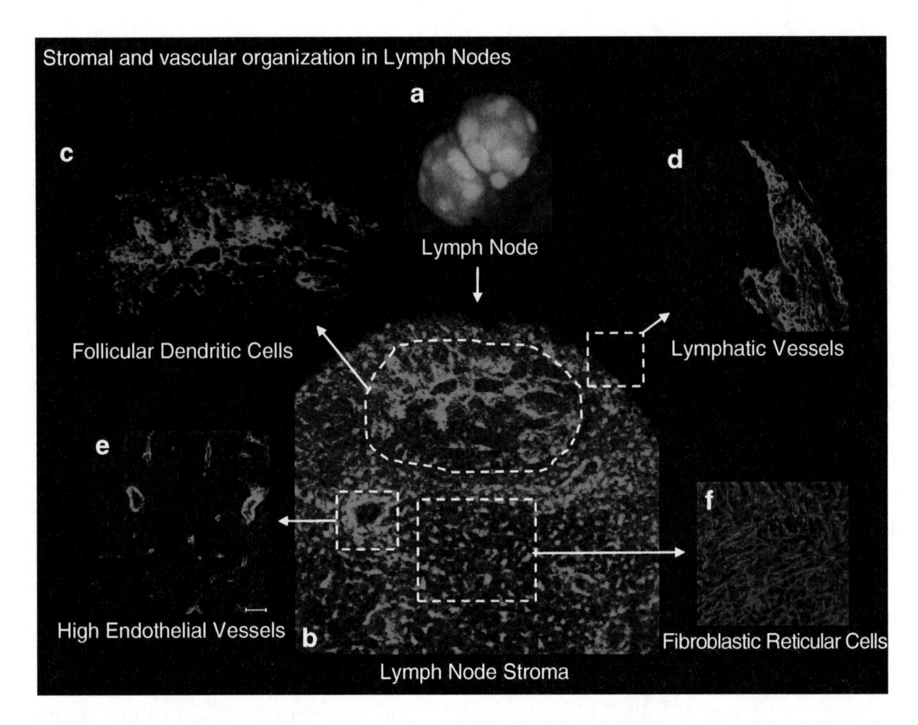


**Fig. 7.2** Stromal and vascular organization of lymph nodes. Lymphocyte and dendritic cells interact with stromal cells that form networks supporting the homeostasis and function of lymphocytes. This highly organized network is supported by vasculature. (**a**) A lymph node from a human CD2.DsRed CD19cre Rosa26eYFP mouse is shown; in these mice, T cells express red fluorescent protein and B cells green. These mice show the distinctive anatomy of lymph nodes. (**b**) Utilizing stromal-specific GFP reporter mice, the underlying stromal network is highlighted. Using antibodies specific for (**c**) follicular dendritic cells (CD35) highlights stromal cells in the B cell follicles, (**d**) lymphatic endothelial cells (Lyve-1) highlight the afferent and efferent lymphatics transporting cells and antigens to the lymph nodes, (**e**) High Endothelial Vessels (Meca79) are found in the T cell zone and are the site of entry for naive lymphocytes into the lymph node, and (**f**) Fibroblastic Reticular Cells (Desmin) form a stromal network that supports T cell migration and interactions with dendritic cells

similar to the mechanism involved in the exit of T cells into the efferent lymphatics. This process is also S1P1 dependent (Schwab and Cyster 2007). Entrance into the medullary cords is dependent on the overcoming of the CCR7 retention signal that acts to attract T cells into the lymph nodes from the HEVs and afferent lymphatics and upregulation of S1P1 required to transmigrate across lymphatic endothelium (Pham et al. 2008). HEVs are only located in the paracortex of the lymph node and as they descend through the peripheral deep cortical units they become progressively larger (Okada et al. 2002). Lymphocytes exit an HEV along its entire length but transmigration is heaviest along the largest HEV segments deep in the paracortex (Gretz et al. 1997; Okada et al. 2002). The transmigration of naïve T cells into lymph nodes from the HEVs like afferent lymphatics is also dependent on CCR7 but unlike lymphatic endothelium it is not dependent on S1P1 signaling.

Thus in PLT mice (Paucity of Lymphocyte T cell), which lack expression of CCL21-Ser and CCL19 entry of naïve T cells into lymph nodes, is inhibited (Luther et al. 2000).

In the murine embryo, HEVs in all the developing lymph nodes express the Mucosal Addressin Cellular Adhesion Molecule-1 (MAdCAM-1) allowing for the selective entry of cells expressing the MAdCAM-1 ligand α4β7 integrin (Mebius et al. 1993; Mebius et al. 1996; Mebius et al. 1998). Postnatally, a dichotomy develops between the mucosal and peripheral lymphoid organs. Where the mucosal-associated HEVs retain expression of MAdCAM-1, HEVs in peripheral lymph nodes start to express Peripheral Node Addressins (PNAd), a collection of sugar moieties recognized by the homing receptor L-Selectin (CD62L). In parallel with PNAd induction, MAdCAM-1 expression starts to decline and from approximately 2 weeks after birth PNAd is the sole addressin expressed on peripheral HEVs (Mebius et al. 1996). This diversification has obvious implications for the regulation of entry into the different lymphoid organs. In utero, LTi cells express the α4β7 integrin, and the expression of MAdCAM-1 on all HEVs allows for the selective recruitment of these cells to the developing lymph nodes (Mebius et al. 1996; Mebius et al. 1998). After entry into the lymph nodes from the blood vasculature, naïve T cells upregulate S1P1 (Pham et al. 2008) allowing egress from lymph nodes. In contrast, antigen-mediated T cell activation leads to CD69 upregulation which binds directly to S1P1 inhibiting cell surface expression of functional S1P1, retaining activated T cells within the lymph node microenvironment (Shiow et al. 2006). Further details on the developmental aspects of lymphocyte homing via HEVs are presented in Chap. 8.

In addition to mediating APC and lymphocyte migration into the lymph node, afferent lymphatic vessels also transport antigen, chemokines, cytokines, and small molecules to the lymph node. This process is mediated by macrophages, which actively phagocytose and transport antigen into the paracortex of the lymph node and specialized conduits that transport small molecules and antigen within the paracortex of the lymph node (Roozendaal et al. 2009). Antigen can enter lymph node through several different active and passive mechanisms, although during pathogen challenge, it is highly likely all of these different mechanisms are involved to differing extents. The differential role of how antigen enters the lymph node during an immune response depends on the degree of innate stimulation, the structural organization of the site of infection, the timing in the immune response, and the type of pathogen. One of the primary mechanisms of antigen entry into the lymph node is through the trafficking of APC from the site of the infection to the draining lymph nodes. The major APC involved in initiating T cell responses are DCs, which actively migrate in a CCR7-dependent manner from tissues through the draining lymphatics to the lymph node localizing in the paracortex proximal to the HEVs where naïve T cells enter (Luther et al. 2000; Forster et al. 2008). The entry of DCs is also dependent on S1P1 and the L1 protein for transmigration across the afferent lymphatic sinus (Czeloth et al. 2005; Maddaluno et al. 2009). A number of other APCs migrate to draining lymph nodes including Langerhans cells and macrophages (Villablanca and Mora 2008; Kirby et al. 2009).

The capacity of lung resident macrophages to transport antigen from the lung to the draining lymph node has been analyzed in *Streptococcus pneumoniae* infection; alveolar macrophages have been shown to very rapidly transport antigen from the lung to the medistinal lymph node (Kirby et al. 2009). The role of macrophages in active transport is unclear; however, the timing is very different with macrophage migration being very rapid and CCR7 independent compared with the CCR7-dependent migration of DCs, potentially implying a role in T independent antibody responses which are important in the clearance of streptococcus infections (Kirby et al. 2009).

Second mechanism of active antigen movement into the lymph node is through the engulfment of antigen complexes by SCS macrophages that line and survey the draining lymphatics. The SCS CD169hi macrophages transport antigen complexes from the afferent lymphatics for presentation to B cells. Unlike conventional macrophages, the SCS macrophages retain opsonized antigen complexes on their cell surface which stimulate B cell activation, B cell uptake of antigen and to rapidly transfer of antigen to FDCs by antigen nonspecific B cells which in turn stimulates efficient B cell responses (Carrasco and Batista 2007; Junt et al. 2007; Phan et al. 2007). Although this is an important mechanism of rapidly stimulating primary and secondary antibody responses to antigen, SCS macrophages are not an important stimulator of T cell responses.

Antigen can be transported into the lymph node through several passive (non-cellular) routes. Very small antigens can enter the lymph node through small gaps in lymphatic endothelium (subcapsular sinus pores), rapidly entering the lymph node from the lymphatics and rapidly stimulating B cells through direct access to B cell follicles. This stimulation was observed within 1 h of infection of GFP-HEL protein with uptake by antigen-specific B cells closest to the SCS (Pape et al. 2007). Within lymph nodes, follicular conduits exist in both the B cell follicles and T cell zone. For some smaller antigens, conduits have been shown to rapidly transport antigen to FDCs in a B cell-independent manner (Roozendaal et al. 2009). A more extensive network of conduits has been shown to connect the SCS to the HEVs in the cortex. These conduits consist of highly organized collagen bundles ensheathed by FRCs and contain a channel permitting movement of small molecules smaller than 70KD (Gretz et al. 2000). These conduits are associated with a population of DCs that are positioned to permit the uptake and rapid presentation of small antigens to T cells (Sixt et al. 2005).

Conduits may also have important role in the transport of chemokines, cytokines, and growth factors into the cortex of the lymph node and B cell follicle; upon antigen stimulation, increased levels of CXCL13 have been observed that perhaps have a role in facilitating the uptake of antigen in the lymph node (Roozendaal et al. 2009). A number of chemokines and cytokines are produced by LECs and may have an important role in stimulating immune responses; however, the role of conduits in the distribution of these factors in the lymph node is unknown. Although movement of antigen by DCs is the prototypic mechanism of how antigen stimulates immune responses, the direct transport of antigen through the lymphatics rapidly drives the initiation of adaptive immune responses long before antigen-specific DCs migrate into the lymph node and initiate the T-dependent immune response (Roozendaal et al. 2009). During an immune responses to pathogen, antigens from the pathogen may enter the lymph node and stimulate response through both active cell migration and passive movement of smaller antigens resulting from pathogen-mediated destruction at the site of the infection.

## References

Adachi S, Yoshida H, Honda K, Maki K, Saijo K, Ikuta K, Saito T, Nishikawa SI (1998) Essential role of IL-7 receptor alpha in the formation of Peyer's patch anlage. Int Immunol 10:1–6

Alimzhanov MB, Kuprash DV, Kosco-Vilbois MH, Luz A, Turetskaya RL, Tarakhovsky A, Rajewsky K, Nedospasov SA, Pfeffer K (1997) Abnormal development of secondary lymphoid tissues in lymphotoxin beta-deficient mice. Proc Natl Acad Sci USA 94:9302–9307

Allen CDC, Okada T, Cyster JG (2007) Germinal-center organization and cellular dynamics. Immunity 27:190–202

Ame-Thomas P, Maby-El Hajjami H, Monvoisin C, Jean R, Monnier D, Caulet-Maugendre S, Guillaudeux T, Lamy T, Fest T, Tarte K (2007) Human mesenchymal stem cells isolated from bone marrow and lymphoid organs support tumor B-cell growth: role of stromal cells in follicular lymphoma pathogenesis. Blood 109:693–702

Ansel KM, Ngo VN, Hyman PL, Luther SA, Förster R, Sedgwick J, Browning JL, Lipp M, Cyster JG (2000) A chemokine-driven positive feedback loop organizes lymphoid follicles. Nature 406:309–314

Bajénoff M, Egen JG, Koo LY, Laugier Jean P, Brau F, Glaichenhaus N, Germain RN (2006) Stromal cell networks regulate lymphocyte entry, migration, and territoriality in lymph NODES. Immunity 25:989–1001

Banks TA, Rouse BT, Kerley MK, Blair PJ, Godfrey VL, Kuklin NA, Bouley DM, Thomas J, Kanangat S, Mucenski ML (1995) Lymphotoxin-alpha-deficient mice. Effects on secondary lymphoid organ development and humoral immune responsiveness. J Immunol 155:1685–1693

Boos MD, Yokota Y, Eberl G, Kee BL (2007) Mature natural killer cell and lymphoid tissue-inducing cell development requires Id2-mediated suppression of E protein activity. J Exp Med 204:1119–1130

Carrasco YR, Batista FD (2007) B cells acquire particulate antigen in a macrophage-rich area at the boundary between the follicle and the subcapsular sinus of the lymph node. Immunity 27(1):160–171

Coles MC, Veiga-Fernandes H, Foster KE, Norton T, Pagakis SN, Seddon B, Kioussis D (2006) Role of T and NK cells and IL7/IL7r interactions during neonatal maturation of lymph nodes. Proc Natl Acad Sci USA 103:13457–13462

Crivellato E, Vacca A, Ribatti D (2004) Setting the stage: an anatomist's view of the immune system. Trends Immunol 25:210–217

Cupedo T, Crellin NK, Papazian N, Rombouts EJ, Weijer K, Grogan JL, Fibbe WE, Cornelissen JJ, Spits H (2009) Human fetal lymphoid tissue-inducer cells are interleukin 17-producing precursors to RORC+ CD127+ natural killer-like cells. Nat Immunol 10:66–74

Cupedo T, Lund FE, Ngo VN, Randall TD, Jansen W, Greuter MJ, de Waal-Malefyt R, Kraal G, Cyster JG, Mebius RE (2004) Initiation of cellular organization in lymph nodes is regulated by non-B cell-derived signals and is not dependent on CXC chemokine ligand 13. J Immunol 173:4889–4896

Czeloth N, Bernhardt G, Hofmann F, Genth H, Forster R (2005) Sphingosine-1-phosphate mediates migration of mature dendritic cells. J Immunol 175:2960–2967

De Togni P, Goellner J, Ruddle NH, Streeter PR, Fick A, Mariathasan S, Smith SC, Carlson R, Shornick LP, Strauss-Schoenberger J et al (1994) Abnormal development of peripheral lymphoid organs in mice deficient in lymphotoxin. Science 264:703–707

Dudda JC, Martin SF (2004) Tissue targeting of T cells by DCs and microenvironments. Trends Immunol 25:417–421

Eberl G, Marmon S, Sunshine MJ, Rennert PD, Choi Y, Littman DR (2004) An essential function for the nuclear receptor RORgamma(t) in the generation of fetal lymphoid tissue inducer cells. Nat Immunol 5:64–73

Endres R, Alimzhanov MB, Plitz T, Fütterer A, Kosco-Vilbois MH, Nedospasov SA, Rajewsky K, Pfeffer K (1999) Mature follicular dendritic cell networks depend on expression of lymphotoxin beta receptor by radioresistant stromal cells and of lymphotoxin beta and tumor necrosis factor by B cells. J Exp Med 189:159–168

Facchetti F, Blanzuoli L, Ungari M, Alebardi O, Vermi W (1998) Lymph node pathology in primary combined immunodeficiency diseases. Springer Semin Immunopathol 19:459–478

Forster R, Davalos-Misslitz AC, Rot A (2008) CCR7 and its ligands: balancing immunity and tolerance. Nat Rev Immunol 8:362–371

Fu YX, Chaplin DD (1999) Development and maturation of secondary lymphoid tissues. Annu Rev Immunol 17:399–433

Fukuyama S, Hiroi T, Yokota Y, Rennert PD, Yanagita M, Kinoshita N, Terawaki S, Shikina T, Yamamoto M, Kurono Y, Kiyono H (2002) Initiation of NALT Organogenesis Is Independent of the IL-7R, LTbR, and NIK Signaling Pathways but Requires the Id2 Gene and $CD3^{-}CD4^{+}CD45^{+}$ Cells. Immunity 17:31–40

Gonzalez M, Mackay F, Browning JL, Kosco-Vilbois MH, Noelle RJ (1998) The sequential role of lymphotoxin and B cells in the development of splenic follicles. J. Exp Med 187:997–1007

Greter M, Hofmann J, Becher B (2009) Neo-lymphoid aggregates in the adult liver can initiate potent cell-mediated immunity. PLoS Biol 7:e1000109

Gretz JE, Anderson AO, Shaw S (1997) Cords, channels, corridors and conduits: critical architectural elements facilitating cell interactions in the lymph node cortex. Immunol Rev 156:11–24

Gretz JE, Norbury CC, Anderson AO, Proudfoot AEI, Shaw S (2000) Lymph-borne chemokines and other low molecular weight molecules reach high endothelial venules via specialized conduits while a functional barrier limits access to the lymphocyte microenvironments in lymph node cortex. J Exp Med 192:1425–1440

Hashi H, Yoshida H, Honda K, Fraser S, Kubo H, Awane M, Takabayashi A, Nakano H, Yamaoka Y, Nishikawa SI (2001) Compartmentalization of peyer's patch anlagen before lymphocyte entry. J Immunol 166:3702–3709

Honda K, Nakano H, Yoshida H, Nishikawa S, Rennert P, Ikuta K, Tamechika M, Yamaguchi K, Fukumoto T, Chiba T, Nishikawa SI (2001) Molecular basis for hematopoietic/mesenchymal interaction during initiation of peyer's patch organogenesis. J Exp Med 193:621–630

Huang JH, Cardenas-Navia LI, Caldwell CC, Plumb TJ, Radu CG, Rocha PN, Wilder T, Bromberg JS, Cronstein BN, Sitkovsky M, Dewhirst MW, Dustin ML (2007) Requirements for T lymphocyte migration in explanted lymph nodes. J Immunol 178:7747–7755

Junt T, Moseman EA, Iannacone M, Massberg S, Lang PA, Boes M, Fink K, Henrickson SE, Shayakhmetov DM, Di Paolo NC, van Rooijen N, Mempel TR, Whelan SP, von Andrian UH (2007) Subcapsular sinus macrophages in lymph nodes clear lymph-borne viruses and present them to antiviral B cells. Nature 450:110–114

Kasajima-Akatsuka N, Maeda K (2006) Development, maturation and subsequent activation of follicular dendritic cells (FDC): immunohistochemical observation of human fetal and adult lymph nodes. Histochem Cell Biol 126:261–273

Katakai T, Hara T, Lee J-H, Gonda H, Sugai M, Shimizu A (2004) A novel reticular stromal structure in lymph node cortex: an immuno-platform for interactions among dendritic cells, T cells and B cells. Int Immunol 16:1133–1142

Katakai T, Hara T, Sugai M, Gonda H, Shimizu A (2004b) Lymph node fibroblastic reticular cells construct the stromal reticulum via contact with lymphocytes. J Exp Med 200:783–795

Katakai T, Suto H, Sugai M, Gonda H, Togawa A, Suematsu S, Ebisuno Y, Katagiri K, Kinashi T, Shimizu A (2008) Organizer-like reticular stromal cell layer common to adult secondary lymphoid organs. J Immunol 181:6189–6200

Kelly KA, Scollay R (1992) Seeding of neonatal lymph nodes by T cells and identification of a novel population of CD3-/CD4+ cells. Eur J Immunol 22:329–334

Kelsoe G (1996) Life and death in germinal centers (redux). Immunity 4:107–111

Kim MY, Gaspal FM, Wiggett HE, McConnell FM, Gulbranson-Judge A, Raykundalia C, Walker LS, Goodall MD, Lane PJ (2003) CD4(+)CD3(-) accessory cells costimulate primed CD4 T cells through OX40 and CD30 at sites where T cells collaborate with B cells. Immunity 18:643–654

Kirby AC, Coles MC, Kaye PM (2009) Alveolar macrophages transport pathogens to lung draining lymph nodes. J Immunol 183:1983–1989

Klein U, Dalla-Favera R (2008) Germinal centres: role in B-cell physiology and malignancy. Nat Rev Immunol 8:22–33

Ledgerwood LG, Lal G, Zhang N, Garin A, Esses SJ, Ginhoux F, Merad M, Peche H, Lira SA, Ding Y, Yang Y, He X, Schuchman EH, Allende ML, Ochando JC, Bromberg JS (2008) The sphingosine 1-phosphate receptor 1 causes tissue retention by inhibiting the entry of peripheral tissue T lymphocytes into afferent lymphatics. Nat Immunol 9:42–53

Link A, Vogt TK, Favre S, Britschgi MR, Acha-Orbea H, Hinz B, Cyster JG, Luther SA (2007) Fibroblastic reticular cells in lymph nodes regulate the homeostasis of naive T cells. Nat Immunol 8:1255–1265

Luther SA, Ansel KM, Cyster JG (2003) Overlapping roles of CXCL13, interleukin 7 receptor {alpha}, and CCR7 ligands in lymph node development. J Exp Med 197:1191–1198

Luther SA, Tang HL, Hyman PL, Farr AG, Cyster JG (2000) Coexpression of the chemokines ELC and SLC by T zone stromal cells and deletion of the ELC gene in the plt/plt mouse. Proc Natl Acad Sci USA 97:12694–12699

Maddaluno L, Verbrugge SE, Martinoli C, Matteoli G, Chiavelli A, Zeng Y, Williams ED, Rescigno M, Cavallaro U (2009) The adhesion molecule L1 regulates transendothelial migration and trafficking of dendritic cells. J Exp Med 206:623–635

Mebius RE (2003) Organogenesis of lymphoid tissues. Nat Rev Immunol 3:292–303

Mebius RE, Breve J, Kraal G, Streeter PR (1993) Developmental regulation of vascular addressin expression: a possible role for site-associated environments. Int Immunol 5:443–449

Mebius RE, Miyamoto T, Christensen J, Domen J, Cupedo T, Weissman IL, Akashi K (2001) The fetal liver counterpart of adult common lymphoid progenitors gives rise to all lymphoid lineages, CD45+CD4+CD3- cells, as well as macrophages. J Immunol 166:6593–6601

Mebius RE, Rennert P, Weissman IL (1997) Developing lymph nodes collect CD4+CD3-LTb+ cells that can differentiate to APC, NK cells, and follicular cells but not T or B cells. Immunity 7:493–504

Mebius RE, Schadee-Eestermans IL, Weissman IL (1998) MAdCAM-1 dependent colonization of developing lymph nodes involves a unique subset of CD4+CD3- hematolymphoid cells. Cell Adhes Commun 6:97–103

Mebius RE, Streeter PR, Michie S, Butcher EC, Weissman IL (1996) A developmental switch in lymphocyte homing receptor and endothelial vascular addressin expression regulates lymphocyte homing and permits CD4+ CD3- cells to colonize lymph nodes. Proc Natl Acad Sci USA 93:11019–11024

Mora JR, von Andrian UH (2006) T-cell homing specificity and plasticity: new concepts and future challenges. Trends Immunol 27:235–243

Mueller SN, Germain RN (2009) Stromal cell contributions to the homeostasis and functionality of the immune system. Nat Rev Immunol 9:618–629

Mueller SN, Hosiawa-Meagher KA, Konieczny BT, Sullivan BM, Bachmann MF, Locksley RM, Ahmed R, Matloubian M (2007) Regulation of homeostatic chemokine expression and cell trafficking during immune responses. Science 317:670–674

Ohl L, Henning G, Krautwald S, Lipp M, Hardtke S, Bernhardt G, Pabst O, Forster R (2003) Cooperating mechanisms of CXCR5 and CCR7 in development and organization of secondary lymphoid organs. J Exp Med 197:1199–1204

Okada T, Ngo VN, Ekland EH, Forster R, Lipp M, Littman DR, Cyster JG (2002) Chemokine requirements for B cell entry to lymph nodes and Peyer's patches. J Exp Med 196:65–75

Pahal GS, Jauniaux E, Kinnon C, Thrasher AJ, Rodeck CH (2000) Normal development of human fetal hematopoiesis between eight and seventeen weeks' gestation. Am J Obstet Gynecol 183:1029–1034

Pape KA, Catron DM, Itano AA, Jenkins MK (2007) The humoral immune response is initiated in lymph nodes by B cells that acquire soluble antigen directly in the follicles. Immunity 26:491–502

Park SY, Saijo K, Takahashi T, Osawa M, Arase H, Hirayama N, Miyake K, Nakauchi H, Shirasawa T, Saito T (1995) Developmental defects of lymphoid cells in Jak3 kinase-deficient mice. Immunity 3:771–782

Pham TH, Okada T, Matloubian M, Lo CG, Cyster JG (2008) S1P1 receptor signaling overrides retention mediated by G alpha i-coupled receptors to promote T cell egress. Immunity 28:122–133

Phan TG, Grigorova I, Okada T, Cyster JG (2007) Subcapsular encounter and complement-dependent transport of immune complexes by lymph node B cells. Nat Immunol 8:992–1000

Reif K, Ekland EH, Ohl L, Nakano H, Lipp M, Forster R, Cyster JG (2002) Balanced responsiveness to chemoattractants from adjacent zones determines B-cell position. Nature 416:94–99

Rennert PD, Browning JL, Mebius R, Mackay F, Hochman PS (1996) Surface lymphotoxin alpha/beta complex is required for the development of peripheral lymphoid organs. J Exp Med 184:1999–2006

Roozendaal R, Mempel TR, Pitcher LA, Gonzalez SF, Verschoor A, Mebius RE, von Andrian UH, Carroll MC (2009) Conduits mediate transport of low-molecular-weight antigen to lymph node follicles. Immunity 30:264–276

Scandella E, Bolinger B, Lattmann E, Miller S, Favre S, Littman DR, Finke D, Luther SA, Junt T, Ludewig B (2008) Restoration of lymphoid organ integrity through the interaction of lymphoid tissue-inducer cells with stroma of the T cell zone. Nat Immunol 9:667–675

Schmidlin H, Diehl SA, Blom B (2009) New insights into the regulation of human B-cell differentiation. Trends Immunol 30:277–285

Schmutz S, Bosco N, Chappaz S, Boyman O, Acha-Orbea H, Ceredig R, Rolink AG, Finke D (2009) Cutting edge: IL-7 regulates the peripheral pool of adult ROR{gamma}+ lymphoid tissue inducer cells. J Immunol 183:2217–2221

Schwab SR, Cyster JG (2007) Finding a way out: lymphocyte egress from lymphoid organs. Nat Immunol 8:1295–1301

Shiow LR, Rosen DB, Brdickova N, Xu Y, An J, Lanier LL, Cyster JG, Matloubian M (2006) CD69 acts downstream of interferon-[alpha]/[beta] to inhibit S1P1 and lymphocyte egress from lymphoid organs. Nature 440:540–544

Sixt M, Kanazawa N, Selg M, Samson T, Roos G, Reinhardt DP, Pabst R, Lutz MB, Sorokin L (2005) The conduit system transports soluble antigens from the afferent lymph to resident dendritic cells in the T cell area of the lymph node. Immunity 22:19–29

Sun Z, Unutmaz D, Zou YR, Sunshine MJ, Pierani A, Brenner-Morton S, Mebius RE, Littman DR (2000) Requirement for RORgamma in thymocyte survival and lymphoid organ development. Science 288:2369–2373

Tsuji M, Suzuki K, Kitamura H, Maruya M, Kinoshita K, Ivanov II, Itoh K, Littman DR, Fagarasan S (2008) Requirement for lymphoid tissue-inducer cells in isolated follicle formation and T cell-independent immunoglobulin A generation in the gut. Immunity 29:261–271

van de Pavert SA, Olivier BJ, Goverse G, Vondenhoff MF, Greuter M, Beke P, Kusser K, Hopken UE, Lipp M, Niederreither K, Blomhoff R, Sitnik K, Agace WW, Randall TD, de Jonge WJ, Mebius RE (2009) Chemokine CXCL13 is essential for lymph node initiation and is induced by retinoic acid and neuronal stimulation. Nat Immunol 10:1193–1199 van Nierop K, de Groot C (2002) Human follicular dendritic cells: function, origin and development. Semin Immunol 14:251–257

Villablanca EJ, Mora JR (2008) A two-step model for Langerhans cell migration to skin-draining LN. Eur J Immunol 38:2975–2980

Vondenhoff MF, Greuter M, Goverse G, Elewaut D, Dewint P, Ware CF, Hoorweg K, Kraal G, Mebius RE (2009a) LT{beta}R signaling induces cytokine expression and up-regulates lymphangiogenic factors in lymph node anlagen. J Immunol 182:5439–5445

Vondenhoff MF, van de Pavert SA, Dillard ME, Greuter M, Goverse G, Oliver G, Mebius RE (2009b) Lymph sacs are not required for the initiation of lymph node formation. Development 136:29–34

Westerga J, Timens W (1989) Immunohistological analysis of human fetal lymph nodes. Scand J Immunol 29:103–112

White A, Carragher D, Parnell S, Msaki A, Perkins N, Lane P, Jenkinson E, Anderson G, Caamano JH (2007) Lymphotoxin a-dependent and -independent signals regulate stromal organizer cell homeostasis during lymph node organogenesis. Blood 110:1950–1959

Wigle JT, Oliver G (1999) Prox1 function is required for the development of the murine lymphatic system. Cell 98:769–778

Yoshida H, Honda K, Shinkura R, Adachi S, Nishikawa S, Maki K, Ikuta K, Nishikawa SI (1999) IL-7 receptor alpha+ CD3(-) cells in the embryonic intestine induces the organizing center of Peyer's patches. Int Immunol 11:643–655

Yoshida H, Naito A, Inoue J, Satoh M, Santee-Cooper SM, Ware CF, Togawa A, Nishikawa S (2002) Different cytokines induce surface lymphotoxin-alphabeta on IL-7 receptor-alpha cells that differentially engender lymph nodes and Peyer's patches. Immunity 17:823–833

# Chapter 8
# Development of Lymph Node Circulation and Homing Mechanisms

**Ann Ager, Mark C. Coles, and Jens V. Stein**

**Abstract** The blood vasculature provides both oxygen and nutrients to the high energetic environment of lymphoid tissue and provides a portal for entry of lymphocytes. The vasculature is uniquely different from that found in other organs due to the requirement for efficient recruitment of lymphocytes under non-inflammatory, physiological conditions. It is highly spatially organised, controlling both the site of cellular entry and contributing to the structural organisation of lymphoid tissues. Understanding the mechanisms of blood vasculature development and function is central to understanding how lymphoid tissues form and to understanding mechanisms regulating adaptive immune responses and disease pathologies. Here, we review the development, maintenance and plasticity of lymphoid blood vasculature, as well as the tissue-specific expression patterns of high endothelial venules, which govern lymphocyte trafficking in steady state and during inflammation.

## 8.1 Overview of Blood Vasculature in Lymphoid Tissues

Blood vasculature is required for both the development and function of secondary lymphoid organs. These specialized tissues include lymph nodes (LNs), Peyer's patches (PPs), tonsils, appendix, bronchus, nasal associate lymphoid tissue (NALT), and tertiary lymphoid organs (TLOs), which arise as a result of chronic

A. Ager (✉)
Department of Infection, Immunity and Biochemistry, School of Medicine, Cardiff University, Cardiff, UK
e-mail: AgerA@cardiff.ac.uk

M.C. Coles
Department of Biology and HYMS, University of York, York, UK
e-mail: mc542@york.ac.uk

J.V. Stein
Theodor Kocher Institute, University of Bern, Bern, Switzerland
e-mail: jstein@tki.unite.ch

P. Balogh (ed.), *Developmental Biology of Peripheral Lymphoid Organs*,
DOI 10.1007/978-3-642-14429-5_8, 

inflammation. All lymphoid organs are highly vascularized and develop specialized vasculature unique to these tissues. The blood vasculature provides both the oxygen and nutrients to the high energetic environment and provides a portal for entry of lymphocytes. The vasculature is uniquely different from that found in other organs due to the requirement for efficient recruitment of lymphocytes under noninflammatory, physiological conditions. It is highly spatially organized, controlling both the site of cellular entry and contributing to the structural organization of lymphoid tissues. In other organs, the nascent vascular network has been shown to have an integral role in their development and organization; thus, it is likely to have a similar role in lymphoid tissues. Understanding the mechanisms of blood vasculature development and function is central to understanding how lymphoid tissues form and to understanding mechanisms regulating adaptive immune responses and disease pathologies.

## 8.2 Vasculature Organization and Development in Lymphoid Tissues

Lymphoid tissue can form during embryonic development (LNs, tonsils, and PPs), as a result of secondary inflammatory stimulation driven either by localized infection (appendix, bronchus, NALT) or through autoimmune stimulation (tertiary lymphoid organs = TLOs). Organized vascular networks and the formation of specialized high endothelial venules (HEVs) are hallmarks of all lymphoid organs, except the spleen. Although the mechanisms driving the development of LNs and PPs are well characterized, the mechanisms driving the formation of TLOs are less clear as they are associated with existing disease pathologies. During embryogenesis, the development of LNs and PPs is regulated by hematopoietic lymphoid tissue inducing cells (LTi) and mesenchymal LTo = lymphoid tissue organising cells. Molecular interactions include TNF superfamily members (lymphotoxin β (LTβ)/LTβ Receptor and RANK/RANK ligand), chemokines/chemokine receptors, IL-7/IL-7 receptor, and adhesion molecules (reviewed in detail in Chaps. 7 and 9). However, studies of interactions between LTi and LTo cells have ignored the potential role of blood vasculature in the formation and organization of LNs. As early as the beginning of the twentieth century, Florence Sabin's study of early stages of human and pig embryo development noted the importance of blood vasculature in lymphoid tissue formation; "When the primary sacs are thus completely bridged by these bands, they are practically a dense plexus of lymphatic capillaries and are therefore in the first stage of the development of lymph glands. At this stage the connective tissue septa are undifferentiated and contain only mesenchyme and blood capillaries". Concerning the subsequent organization of lymphoid tissues; "There are two processes in the development of the follicle (1) an increase in the number of lymphocytes forming a definite clump and (2) the formation of a tuft of blood capillaries. Both the cords and the follicles form along the blood vessels, the follicles coming at the capillary bed"

(Sabin 1913). Thus, the correlation between blood vasculature development and the organization and formation of the anlagen has long been associated. Indeed, the paradigm that vasculature development is important in organogenesis has been demonstrated in other organs. During pancreas development, the developing blood vasculature is the source of signals required for the induction of insulin production by the endoderm (Lammert et al. 2001). Analysis of lung development has shown an essential role for Ang1 and VEGFs in regulating the development of the normal lung (van Tuyl et al. 2007). Vascularization is found early in the development of other lymphoid organs. In the thymus, we also observe blood vasculature development prior to e12.5 consistent with vascularization being important in early thymus organogenesis (Foster et al. 2008).

In LNs, the blood vasculature is highly organized. The transfer of nutrients and oxygen occurs efficiently via the extensive capillary bed surrounding B cell follicles and in the T cell zone. Efficient lymphocyte entry occurs from specialized HEVs in the deep cortex of the LN. Soluble antigen (Gretz et al. 2000) and dendritic cells (DCs) (Randolph et al. 2005) enter the LN through the afferent lymphatics. Low molecular weight antigens are transported via FRC = fibroblastic reticular cells ensheathed, type I collagen containing conduits, permitting efficient sampling by DCs lining the conduits and subsequent presentation to incoming T lymphocytes (Sixt et al. 2005). Arteries feeding the LN enter at the hilar region and arborize into a capillary bed in the outer cortex, which leads directly (or indirectly via arteriovenous shunts) into the postcapillary venular network. This traverses the paracortex (T cell area) of the LN, and HEVs are located within this postcapillary network. Several major HEV trunks, each receiving 3–5 smaller side-branches as they traverse the paracortex from cortex to medulla, are found within each LN. HEVs have been identified according to their location from smallest branches at the inner cortical/paracortical junction (order V) through to order III vessels just prior to the largest vessels in the paracortical/medullary junction (lower order II) (Warnock et al. 1998). In the medulla, order II vessels merge into low order I collecting venules, which drain into hilar vein (Fig. 8.1). The endothelial cells lining HEV form a sharp transition from flattened to plump morphology at the capillary boundary with order V venules (Anderson and Shaw 1993). The majority of HECs exhibit a basal level of activity, but are interspersed with cells of increased cytoplasmic density due to abundant polyribosomes, dilated Golgi and an invaginated luminal surface. This heterogeneity in metabolic activity is accompanied by varying levels of nonspecific esterase and perhaps hot spots of enhanced lymphocyte extravasation. The unique structure of HEV allows lymphocytes to transmigrate the vessel wall without breakdown in vascular integrity, as found in inflamed vessels outside of LNs, which are permeable to macromolecules.

The actual height of the endothelial lining varies between species, and according to the method of tissue collection, therefore these vessels cannot be defined on morphological terms alone. For example, endothelial lining of equivalent vessels in sheep, rabbits, and guinea pigs is not as high. Collection of rodent LNs under reduced arterial pressure results in constriction of the vessel lumen and plumping up of the endothelium to 15–20 μm height (Belisle and Sainte-Marie 1985). If the

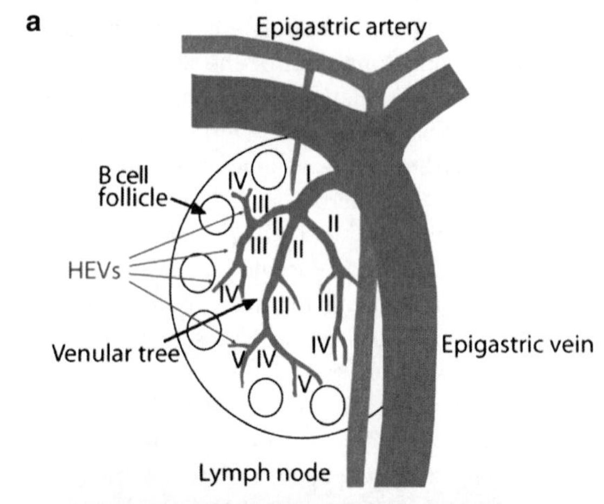


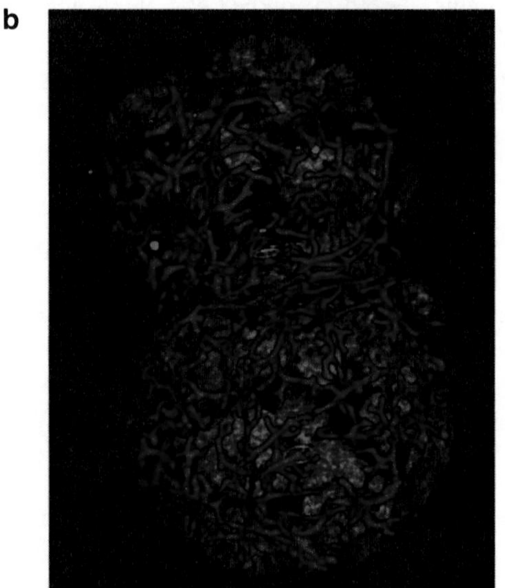


**Fig. 8.1** (**a**) Schematic representation of the microvascular network of a mouse inguinal LN. The vessel order starts from the largest collecting vein (order I) and ends in the smallest immediate postcapillary venules (order V). HEVs are formed by vessels within orders II–V and localize in the T cell area surrounding B cell follicles. Adapted from von Andrian 1998. (**b**) 3D reconstruction of a mouse inguinal LN showing B cell follicles in *green* and the HEV network in *red*. Note the extensive combined length of the HEV network responsible for continuous lymphocyte recruitment

patency of HEV is maintained by perfusion during fixation then the endothelium is 5–7 μm in height, although this is still greater than that lining non-HEV blood vessel. The relationship between vessel patency and endothelial height suggests that HEVs are contractile vessels. The endothelial cell and the pericyte could both regulate HEV contractility. Dynamic changes in HEV morphology may enhance the trapping and passage of lymphocytes by altering shear stress at the vessel wall interface.

Postcapillary venules in other tissues are surrounded by a sheath of pericytes, which are embedded in an extracellular matrix of secreted fibronectin, laminin, and collagens. Controversy surrounds whether pericytes also provide coverage to the HEV networks within the LN. This might be part result from localized differences between individual LNs and the close structural and functional similarity between pericytes and FRCs. Indeed, the difference between these two cell types might only be in the functional role of the two cell types and not in their relative gene expression profile. Like pericytes, FRCs express desmin, smooth muscle actin, collagen IV, and other extracellular matrix components. Analysis of pericytes surrounding HEVs in lymphoid tissues shows that they express CCL21. The only phenotypic difference between pericytes and the FRC network is the expression of the chondroitin sulphate proteoglycan NG2 (m.c.c., Unpublished observations). NG2 expression by pericytes is associated with newly formed blood vasculature and thus the continual expression of NG2 in LNs might represent the ongoing plasticity of the vascular network. The very close phenotypic relationship between FRCs and pericytes implies a potential developmental relationship between both cell types. Analysis of stromal networks in TLOs suggests local differentiation of stromal networks, consistent with the idea that FRCs might arise from pericytes at the site of LN development and TLO formation. Pericytes also secrete collagen I, which forms into conduits between the lymphatic endothelium and the HEVs in the developing LN (Fig. 8.2). How the complex bundles of collagen I form into highly

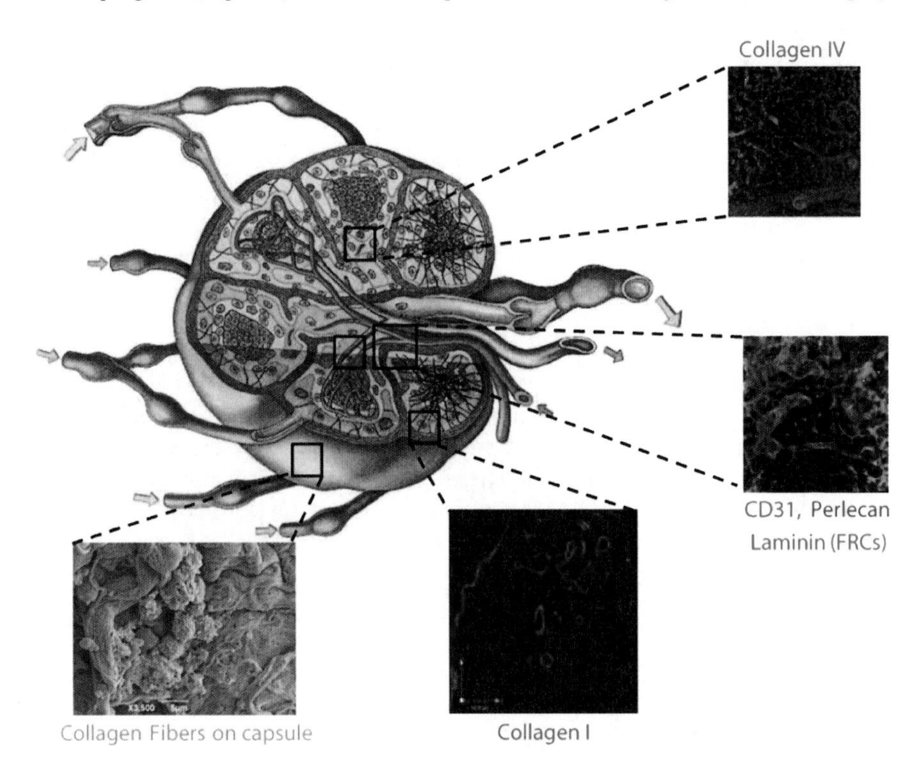


**Fig. 8.2** Collagen networks in LNs. Collagen IV forms part of the basement membrane surrounding blood vessels, while fibrous collagens form the capsule, and like collagen I, form part of the core of the FRC network. Similarly, Perlecan and laminin are part of the inner FRC network core, while CD31 is specific for blood and lymphatic endothelial cells

organized secondary structures is unknown; however, the localization of blood vasculature appears to have a key role in this process as conduits connect the draining lymphatics with the HEVs (Gretz et al. 2000).

## 8.3 Modification and Plasticity of Vascular Networks in LNs During Development

During lymphoid tissue development, the formation of B cell follicles in mice does not occur until after birth as B cells do not start populating LNs until very late stages of development. Formation of the follicles is initiated by B cells in a CXCL13-, TNF-, and LTβ-dependent mechanism. Formation of a capillary network, which surrounds the B cell follicle, appears to occur concurrently with the developing follicle. It is unclear if formation of the vascular network is a consequence of follicle formation, or whether a preexisting vascular bed dictates the organization and location of developing follicles. The formation and rapid remodeling of vascular network in LNs during neonatal development is similar to the process that occurs during immune responses (see below). The nascent HEVs in all lymphoid tissues express mucosal addressin cell adhesion molecule (MAdCAM)-1, which recruits LTi into the developing LN anlagen via LTi-expressed α4β7 integrin (Mebius 2003). As peripheral LNs mature in neonatal pups, they lose expression of MAdCAM-1 and upregulate PNAd = peripheral node addressin expression in an LTβ-dependent manner (Drayton et al. 2006; Mebius et al. 1996). Interestingly, this appears to be dependent on the location and level of antigenic stimulation in the LN microenvironment. In mucosal LNs, both PNAd and MAdCAM-1 are coexpressed, whereas PNAd is absent from PP in mice, but highly expressed in PPs in other species (Rosen 2004). Although the induction and maintenance of HEVs is dependent on continual LTβ signaling, the mechanisms leading to gain of PNAd expression and loss of MAdCAM-1 in peripheral nonmucosal LNs are unknown. Potential mechanisms that might have a role are small molecular factors secreted by LECs = lymphatic endothelial cells. Thus, interruption of the afferent lymph flow in adult mouse LNs (Mebius et al. 1991) results in downmodulation of PNAd expression patterns in endothelial cells and transient upregulation of MAdCAM-1 (see below). Such lymph-borne factors might include retinoic acid, vitamin D3, cytokines, growth factors, or soluble lymphotoxin.

## 8.4 The HEV Phenotype

In rodents and humans, postcapillary endothelial cells lining HEV have a plump or cuboidal morphology, which contrasts with the flattened morphology of endothelial cells lining other types of vessel. It is the endothelial morphology that has engendered the common name of HEV. The function of HEV is to sort lymphocytes from other leucocytes circulating in the bloodstream and to deliver them into the LN for sampling by antigen-loaded DCs. It is, therefore, reasonable to assume that structural and functional properties unique to HEV, which are not shared by other postcapillary venules, regulate lymphocyte recruitment from the bloodstream. Defining properties of HEV include the expression of vascular addressins and arrest chemokines on the luminal inner surface of HEV, which cooperate in selecting lymphocytes for extravasation (see below). The irregular luminal surface of HEV may help generate the shear stresses required for efficient lymphocyte binding.

The recirculation of lymphocytes through lymphoid organs is not random. Several mechanisms cooperate to regulate differential homing of lymphocyte subsets to distinct locations with lymphoid tissues and to different lymphoid tissues. These reflect the combination of adhesion receptors and chemokine receptors used to roll and arrest on the inner HEV surface and the location of complementary adhesion ligands and arrest chemokines within the HEV network. The translocation of lymphocytes from circulating blood across the vessel wall is a rapid event taking as little as 10 min to complete (Smith and Ford 1983) and 1 in 4 lymphocytes passing through HEV are recruited into LNs (Hay and Hobbs 1977). The speed and efficiency of recruitment and the ability to image lymphocyte behavior inside HEV of intact animals has allowed the first stages of recruitment to be studied intensively (Bargatze et al. 1995; Warnock et al. 1998). In common with recruitment of other types of leukocytes, it follows the rules of the adhesion cascade (Fig. 8.3). Lymphocyte recruitment into peripheral LNs occurs largely from higher order venules III–V. Initial contact (tethering) and subsequent reversible adhesion, which manifests as rolling under flow, are mediated by L-selectin-dependent binding to PNAd

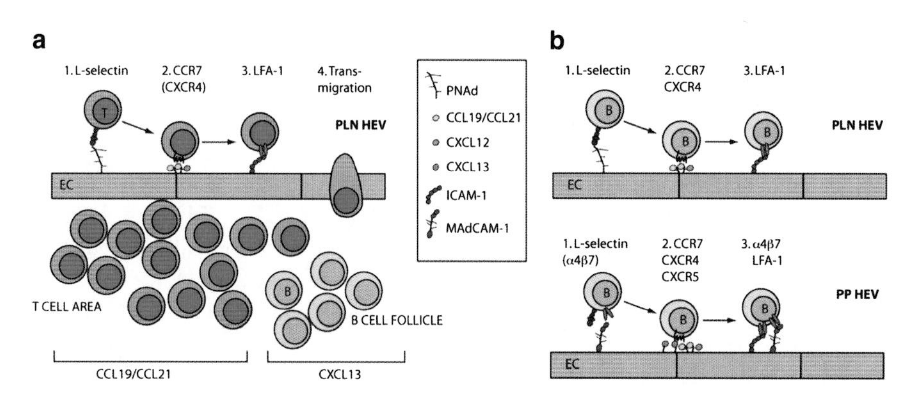


**Fig. 8.3** (**a**) The multistep homing cascade of T cells into peripheral LNs. After undergoing L-selectin-PNAd-mediated rolling, CCL19 and CCL21 presented on the HEC surface activate the CCR7 chemokine receptor on the T cell. This results in an outside-in signaling cascade activating the LFA-1 integrin, binding its endothelial counterligand ICAM-1. After transmigration, localized expression of CCL19 and CCL21 vs. CXCL13 determines the T and B cell areas. (**b**) Multistep B cell homing to peripheral LNs (*upper panel*) and PPs (*lower panel*). In LN HEVs, B cells use CXCR4, in addition to CCR7, for integrin activation. In PP HEV, the α4β7 integrin can mediate rolling and firm adhesion by binding its endothelial ligand MAdCAM-1. In addition, the CXCL13 is found on HEV segments close to B cell follicles

on the luminal surface of HEV (step 1). G-protein coupled chemokine receptor (CCR7) activation by CCL21 immobilized on the endothelial surface activates LFA-1 integrin-mediated arrest on ICAM-1 and/or ICAM-2 (Lehmann et al. 2003; Stein et al. 2000) (step 2). Lymphocytes then penetrate the apical surface and exit from the basolateral surface to transmigrate the endothelial lining (step 3). Finally, lymphocytes penetrate the surrounding pericyte embedded basement membrane to enter the flow-free parenchyma of the LN (step 4). The adhesion receptors, chemokine receptors, and signaling pathways that regulate the translocation of lymphocytes from the luminal surface of HEV into LN interstitium are poorly understood largely because it has been difficult to image this step in real time in intact animals or organs. The location of lymphocytes within the endothelial lining and underlying basement membrane are defining features of these vessels in histological sections, which suggests that these are rate-determining stages (steps 3 and 4). Pulse chase and intravital analysis of intravenously administered lymphocytes shows that it takes 3 min to cross endothelial lining of HEV and 10 min to exit from the basement membrane (Bajenoff et al. 2006). Unresolved issues are whether lymphocytes pass between adjacent EC (paracellular route) or through EC cytoplasm (transcellular); where and how do lymphocytes detect and respond to immobilized chemokines in the vessel wall for directed migration into the LN (Miyasaka and Tanaka 2004); how do lymphocytes exit the vessel wall and enter the LN parenchyma. Interventional studies using CAM antagonists and protease inhibitors as well as intravital imaging techniques using HEV reporter mice (Liao et al. 2007) are starting to dissect these later stages of extravasation.

## 8.5 Tissue-Specific Homing: Peripheral and Mucosal Addressins

This behavior is determined by the expression of "homing receptors" on lymphocytes that bind to tissue-specific adhesion receptors on HEV. These are termed vascular addressins because they provide geographical cues or address codes to circulating lymphocytes (Fig. 8.3). Peripheral LNs and mucosal addressins have been identified in mice using mAb MECA 79 and MECA 367 binding to PNAd and MAdCAM-1, respectively, and are defining features of HEV in LNs and gut-associated lymphoid tissues (Drayton et al. 2006). PNAd is expressed in LNs draining subcutaneous sites such as subiliac/inguinal, axillary, brachial, and popliteal. However, lymphoid tissues draining mucosal sites such as cervical (submandibular) LN or NALT express PNAd and not mucosal addressin (Drayton et al. 2006); thus, addressin expression cannot be used in isolation to define LN drainage area. Interestingly, although PNAd is expressed at the basolateral surface of HEV in PPs, it is not accessible to lymphocytes circulating in the bloodstream (Rosen 2004). However, L-selectin does support tethering and rolling in Peyer's patch HEV via the mucin domain of MAdCAM-1 (Berg et al. 1993), although $\alpha4\beta7$

integrin mediates slower rolling on MAdCAM-1 (Berlin et al. 1993). α4β7 integrin also cooperates with LFA-1 to mediate chemokine receptor-mediated arrest and can compensate for lack of LFA-1. Peripheral LN homing and cellularity are most affected in LFA-1 deficient mice, while residual homing is mediated by α4 integrins (VLA-4>α4β7) binding to VCAM-1 expressed on the luminal surface of HEV (Berlin-Rufenach et al. 1999). α4 integrins may therefore bridge L-selectin-dependent fast rolling and LFA-1-dependent arrest during steady state homing in nonmucosal as well as in mucosal LNs.

## 8.6 Lymphocyte Subset Homing Mechanisms

In comparison with PPs, peripheral LNs are enriched in T cells and depleted in B cells, and this nonrandom distribution of lymphocytes is partly regulated at the level of recruitment from the bloodstream. B cells express low levels of L-selectin at ~50% of that on T cells. A direct comparison with T cells expressing equivalent low surface levels of L-selectin showed reduced recruitment to peripheral LNs (Tang et al. 1998). Lymphocytes expressing reduced levels of L-selectin are particularly sensitive to alterations in PNAd expression. Thus, HEVs expressing glycan-deficient PNAd will select T cells over B cells on the basis of high L-selectin expression (Gauguet et al. 2004). Naïve and central memory T cells use L-selectin and CCR7 to roll and arrest on HEV in peripheral LNs. However, in the absence of CCR7, CXCR4 can mediate arrest of memory, but not naïve, T cells via CXCL12 presented on HEV surface (Scimone et al. 2004). CXCR4 can also regulate B cell recruitment to peripheral LNs as well as PPs, although CCR7 predominates (Okada et al. 2002). There is also variation within the HEV network, which imparts segmental or regional control over lymphocyte homing. In PPs, CXCL13 is presented on the luminal surface of HEV at the inner cortical/paracortical junction (order V vessels) and arrests B cells, whereas T cells arrest in response to CCL21 expressed in lower order III and IV venules (Okada et al. 2002; Warnock et al. 2000). This demonstrates that B cell recruitment occurs preferentially in the cortical/paracortical junction for subsequent localization in the B-cell enriched outer cortex of PP. Segmental control of homing may also be regulated by vascular addressin composition since the structure of PNAd differs between small and large HEV (see below).

## 8.7 How Do Lymphocytes Transmigrate and Exit HEV?

Since the suggestion by Marchesi and Gowans that lymphocytes penetrated the endothelial cytoplasm (transcellular route) rather than migrating between adjacent endothelial cells (paracellular route), the actual route of migration across the vessel wall has been a matter of debate (Engelhardt and Wolburg 2004; Marchesi and

Gowans 1964). Although lymphocytes are routinely found completely engulfed within the cytoplasmic body of individual endothelial cells, evidence in support of both routes of transmigration has been presented. In fact, one report concluded that lymphocytes passed between adjacent endothelial cells after an initial passage through the luminal portions of endothelial cytoplasm. Scanning electron microscopic analysis of the luminal surface of HEV shows that high endothelial cells (HECs) form a continuous lining, and there are no gaps between HECs or within HEC cytoplasm for lymphocytes to penetrate (Anderson and Shaw 1993). Depending on the location in the HEV network, HECs are either arranged in a cobblestone array (proximal) or as an interlacing network of cells (distal segment). All HECs have cytoplasmic extensions, which overlap with adjacent HECs. There are deep crevices in the luminal surface at these intercellular contacts and lymphocytes adhere preferentially at or near these junctions. But what drives directed movement from the apical to basolateral surface of HECs is not clear.

Endothelial cells lining HEV express a range of junctional proteins, which are found in other vascular beds, including VE-cadherin, CD31, JAM-A, JAM-B, JAM-C, and ESAM-1 (Pfeiffer et al. 2008). The sequential engagement of junctional proteins located within the inter-EC junction by complementary receptors on lymphocytes may direct transendothelial migration and has been found for leucocytes transmigrating inflamed blood vessels (Vestweber 2007). HEV lack structurally distinct tight junctions and vascular-specific claudin-5, which may facilitate the paracellular route of lymphocyte transmigration.

The molecular basis of transcellular migration is not worked out, but docking structures enriched in ICAM-1 and VCAM-1, which project from the luminal surface of cultured EC, have been reported following adhesion arrest in vitro (Barreiro et al. 2002; Carman and Springer 2008). Lymphocytes form very close, intercellular, gap-like junctions of 2–4 nm with EC during transmigration across HEV (Campbell 1983), which may be the in vivo equivalent of the docking structures reported in vitro. Interestingly, in mice treated systemically with ADAM inhibitors, T cells accumulated within the endothelial lining of HEV and did not enter the LN parenchyma, indicating that exit from the basolateral surface, rather than penetration of the apical surface, is metalloproteinase dependent (Faveeuw et al. 2001). L-selectin undergoes ADAM-dependent ectodomain cleavage and is down-regulated from T cells as they transmigrate HEV (Klinger et al. 2009). Interestingly, T cells expressing a non-cleavable mutant of L-selectin take longer to transmigrate HEV and move into the LN parenchyma than wild type T cells (Galkina et al. 2003). L-selectin shedding may be required to regulate signaling via PNAd expressed at high endothelial cell surfaces which, in turn, affects lymphocyte motility (Harris and Miyasaka 1995).

The endothelial lining is surrounded by a basement membrane comprising a layer of pericytes embedded in mixture of extracellular matrix proteins typical of basement membranes including laminin, fibronectin, glycosaminoglyans, and type IV collagen (see above). The basement membrane, but not the endothelial lining, is a barrier to macromolecules in the bloodstream. Following intravenous administration, electron dense macromolecules accumulate on the endothelial side of the basement membrane (Anderson and Shaw 1993). MECA 79, a 750 kD IgM, which normally stains the luminal surface of HEV following i.v. infusion, will also stain PNAd at the basolateral surfaces of HEV if administered under increased arterial pressure (Streeter et al. 1988). Conversely, molecules entering the LN via the afferent lymphatics have access to the intravascular compartment. The FRC ensheathed conduits mediate the bulk flow of soluble chemoattractants from the incoming, afferent lymphatics to HEVs and presentation of inner, luminal surface of HEV by reverse transcytosis (Baekkevold et al. 2001).

## 8.8 HEV-Specific Genes

### *8.8.1 Peripheral LN Addressin, L-Selectin Ligand Sulphotransferase and GlyCAM-1*

PNAd, identified by mAb MECA 79, is a unique marker of HEV in mice, man, and other species [for review see Rosen (2004)]. The MECA 79- and L-selectin-reactive epitopes overlap, and MECA 79 inhibits L-selectin-dependent binding to HEVs. L-selectin-reactive capping groups depend on sulphation, fucosylation, and sialylation and a major capping group is 6-sulpho sialyl Lewis$^{x}$. Sulphation of this epitope is regulated by *N*-acetylglucosamine-6-O-sulphotransferases and one isoform, GlcNAc6ST-2, is restricted to HEV [also known as L-selectin ligand sulphotransferase (LSST) or HEC-6ST] (Hemmerich et al. 2001). HEC-specific GlcNAc6ST-2 generates ligands expressed on the inner luminal surface of HEV, which regulate the rolling velocity of blood-borne lymphocytes (van Zante et al. 2003). HEV also express GlcNAc6ST-1, which cooperates with GlcNAc6ST-2 to generate MECA 79 epitope and shear stress-dependent L-selectin ligand binding (Uchimura et al. 2005). Although GlcNAc6ST-1-dependent MECA 79-reactive ligands are also expressed at the basolateral endothelial surface, these are not accessible to lymphocytes circulating in the bloodstream (Hemmerich et al. 2001). L-selectin-reactive capping groups are generated in both extended core-1- and core-2-branched O-glycans. Although the MECA 79 epitope is restricted to extended core-1 structures within O-glycans expressed by HEV, MECA 79 and L-selectin recognize the same glycoprotein complex in HEV termed PNAd. This includes the CD34 family of sialomucins (CD34, podocalyxin, endogycan), nepmucin (Umemoto et al. 2006), endomucin (Kanda et al. 2004), GlyCAM-1, MAdCAM-1, and, as yet, unidentified proteins such as Sgp200 (Rosen 2004). Of these, GlyCAM-1 is specific to HEV, only being expressed elsewhere in mammary epithelium. Since GlyCAM-1 is secreted into plasma, it is unlikely to function as an adhesive ligand at the HEV surface. As yet, GlyCAM-1 has not been detected in human lymphoid organs. In contrast, although CD34 is expressed widely on vascular endothelia and haemopoietic stem cells, only HEV-expressed CD34 is appropriately modified and functions as a major L-selectin ligand in LN. Other glycosyltransferases necessary to generate PNAd include fucosyltransferase VII, core 2 branching enzyme (Core

GlcNAcT), and core 1 extension enzyme (Core 1-β3GlcNAcT). There is considerable redundancy in use of ligands for L-selectin since deletion of CD34, GlyCAM-1, or MECA79 epitope does not completely abrogate lymphocyte homing in mice. Additional MECA $79^{neg}$ ligands include those modified by fucosyltransferase IV found in larger venules (low order I) (M'Rini et al. 2003), 6-sulpho sialyl Lewis x on N-glycans (Mitoma et al. 2007), E-selectin, and undersulphated sialyl $Lewis^{x}$ (Kawashima et al. 2005; Mebius and Watson 1993).

### 8.8.2 *Arrest Chemokine CCL21*

The pertussis toxin-sensitive arrest of rolling lymphocytes in HEV is mediated by CCL21 (also known as secondary lymphoid chemokine, SLC). Mice express 2 genes, which encode CCL21-ser and CCL21-leu. Of these, CCL21-ser is selectively expressed by HEV endothelial cells and lymphatics in LN and PPs, and CCL21-leu is restricted to lymphatic endothelium in nonlymphoid organs (Nakano and Gunn 2001; Vassileva et al. 1999). The receptor on lymphocyte is CCR7, which plays an important role in recirculation of lymphocytes through LNs (Stein et al. 2000). Other ligands include CCL19 (ELC) and although not synthesized by HECs, it can be presented on the luminal surface following reverse transcytosis from the basolateral surface (Baekkevold et al. 2001).

### 8.8.3 *Other HEV-Expressed Genes Implicated in Lymphocyte Trafficking*

Attempts to identify HEV-specific genes by comparing RNA expression libraries prepared from PNAd expressing HEC and "flat" EC identified HEVin, a SPARC-like matricellular protein secreted by HEC and localized at lateral junctions (Girard and Springer 1995). The potential role of HEVin in lymphocyte trafficking has not been explored but immobilized HEVin decreased adhesion of endothelial cells to ECM = extracellular matrix, and thus HEVin may regulate the distinct morphology of HEV (Girard and Springer 1996). Other gene products, which although not specific to HEV, are increased in comparison with "flat" EC. For example, autotaxin secreted by HEV regulates lymphocyte recruitment from HEV by virtue of its phospholipase D activity, which generates lysophosphatidic acid (LPA) (Kanda et al. 2008). LPA induces lymphocyte chemokinesis and altered morphology in endothelial cells, and thus the effect of autotaxin on lymphocyte recruitment may be mediated via both lymphocytes and endothelial cells (Nakasaki et al. 2008). Additional proteins upregulated in HEV include DARC, IL-33/NF-HEV, mac25, and a leucine-rich region containing protein with 67% homology to human leucine-rich α2-glycoprotein (Carriere et al. 2007; Izawa et al. 1999). An interesting observation is the abundance of protease inhibitors expressed by HEC including α2-macroglobulin, cystatin A, SPINK-5, and α-protease inhibitor (Carriere et al. 2007). However, further studies are required to determine their precise roles in HEV function. For example, although DARC does not regulate constitutive lymphocyte recirculation (Kashiwazaki et al. 2003), it may affect recruitment into inflamed LNs (see below). Similarly, although engagement of Eph receptors on $CD4^+$ T cells upregulate chemokine responses and ephrin A1 is expressed by HEV, a definite role for ephrin receptor signaling in lymphocyte trafficking in vivo remains to be determined (Aasheim et al. 2005).

## 8.9 Maintenance of HEV Phenotype

In adult rodents, the expression of vascular addressins, arrest chemokines, and cuboidal endothelial morphology of HEV are all actively maintained. Early studies showed that ligation of afferent lymphatics the so-called deafferentisation, which drain from the surrounding tissues into popliteal LN, has a dramatic effect on the phenotype and function of HEV (Hendriks et al. 1987; Mebius et al. 1993; Mebius et al. 1991; Swarte et al. 1998). PNAd, GlyCAM-1, and FucTVII expression decrease dramatically over the first week, and luminal expression of PNAd is undetectable after 8 days, although abluminal PNAd expression is still detectable, which could reflect maintained expression of GlcNAc6ST-1 over HEC-6ST/GlcNAc6ST-2. There is a transient increase in MAdCAM-1, which peaks 4 days following deafferentization. Lymphocyte recruitment by HEV decreases in line with falling luminal PNAd expression and is barely detectable after 8 days. There is also a gradual flattening of endothelial cells lining HEV, but the contribution of morphological changes to HEV function is not clear. Following isolation from the lymphoid tissue microenvironment, HECs also rapidly dedifferentiate, showing early loss of FucTVII and a more gradual loss of HEC-6ST, CCL21, and Core1-β3GlcNAcT gene expression (Lacorre et al. 2004), suggesting that some trafficking associated genes may be independently regulated in HEV.

The components of afferent lymph that regulate HEV function are not completely worked out but a clue has come from studies of the role or LTβR signaling in LN development. It has long been known that blood vessels with the morphology and function of HEV develop at extranodal sites of chronic immune-mediated inflammation such as in type I diabetes or rejecting allografts. LTβR signaling via the noncanonical NFκB pathway plays a key role in the induction of HEC-6ST and PNAd at extranodal sites (Drayton et al. 2003). Interestingly, HEV-specific genes HEC-6ST and PNAd, GlyCAM-1, and CCL21 are all targets of $I\kappa\kappa\alpha^{AA}$ critical signaling molecule in the noncanonical NFκB pathway (Drayton et al. 2004). LTβR is expressed by HEV in peripheral LNs, which suggests that LTαβ or alternative LTβR ligands are responsible for maintaining the HEV phenotype. Chronic administration of LTβR blocking reagents to adult mice recapitulates the effects of deafferentization in which HEV selectively lose aspects of its fully differentiated status including most functional PNAd and MAdCAM-1 and therefore its ability to support lymphocyte homing (Browning et al. 2005). The identity and cellular source of ligands in afferent lymph for IKKα signaling receptors on HEV remain to be determined. The systematic administration of defined cellular and soluble components to isolated HECs to maintain a differentiated phenotype will be useful in this analysis.

## 8.10 Dynamic Changes of the Lymphoid Structure During Inflammation

When microbes or, in experimental settings, microbial products or activated DCs enter a host, the draining LNs and other lymphoid tissues typically undergo profound changes in their size and microenvironmental compartmentalization. Macroscopically, LNs often increase several-fold in size during the time course of a few days and remain swollen for up to several weeks. The cellular composition of LNs also changes, with additional inflammatory cells, including monocytes, NK cells, and neutrophils, entering LNs (Chtanova et al. 2008; Martin-Fontecha et al. 2004; Palframan et al. 2001). Since the immunological requirements for clearing of infections vary according to the challenge, general rules for the restructuring of lymphoid tissue have not yet been established in a comprehensive fashion. Nonetheless, a role for innate immune cells in remodeling of feeding arterioles has been described, resulting in increased blood supply and increased delivery of naïve lymphocytes into draining LNs (Soderberg et al. 2005). This results in an efficient recruitment of a large part of the lymphocyte repertoire through a draining LNs within a few days (Hay and Hobbs 1977). To accommodate the increased blood supply and necessity of lymphocyte trafficking to draining LNs, the HEV network also grows considerably in size. As an example, while a noninflamed mouse inguinal LN contains a total HEV length of approximately 10 cm, this increases to 30 cm during certain viral infections, in parallel with a threefold size increase of the total LN volume (Kumar et al., 2010). Similar to LN development during embryogenesis, members of the TNF family, including LT and TNF, appear to play a central role in this remodeling. Administration of blocking decoy receptor proteins efficiently reduces LN and HEV growth, with LT playing a dominant role in certain experimental settings (Browning et al. 2005). The cellular source of LT ligands is currently unknown, although B cells expressing LTα1β2 are critical for LN remodeling during some but not all proinflammatory settings (Angeli et al. 2006; Halin et al. 2007). In addition, LT ligand-expressing activated T cells may also function as morphogenic factors. Since HEV express the LTβR, direct interactions between transmigrating lymphocytes and endothelial cells of the HEV network may induce growth, although this has not been yet investigated. Furthermore, LTβR activation on FRCs induces expression of the prototypical angiogenic factor VEGF-A, which in turn induces HEV growth, although other angiogenic mechanisms are likely to exist (Chyou et al. 2008). VEGF-A is also expressed by activated DCs, which often localize close to HEV (Mempel et al. 2004; Webster et al. 2006). Finally, in addition to LTα1β2, activated B cells secrete VEGF-A (Angeli et al. 2006). Thus, both members of the TNF and VEGF family exert morphogenic effects on draining LNs. It is likely that inflammation-dependent, yet undescribed regulatory mechanisms induce the individual or concerted activation of these morphogenic pathways, with redundancy between the cellular sources and responders of these factors.

Further to the remodeling of the HEV network and the LN size, the characteristic microenvironmental separation of T and B cell areas often becomes lost at the peak of inflammation (Katakai et al. 2004; Mueller et al. 2007; Scandella et al. 2008). The parenchymal dispersion of T and B cells is paralleled by, and likely due to, a transient loss of homeostatic chemokines, such as CCL21 and CXCL13 (Mueller et al. 2007; Scandella et al. 2008). Furthermore, the expression of PNAd is transiently downregulated, which may be due to an inflammation-induced loss of LTβR expression (Browning et al. 2005; Liao and Ruddle 2006). These observations hint to a negative feedback regulation of LN size, which impairs uncontrolled growth of inflamed lymphoid tissue. Indeed, a largely unknown issue is how the homeostatic structure of LNs and other secondary lymphoid tissues is restored through molecular and cellular mechanisms. A recent study provides evidence for increased recruitment of LTi at later stages in a viral infection model, and this recruitment is critical for restoration of the preinfection architecture of lymphoid tissues, thus recapitulating the events during LN development (Scandella et al. 2008). Future studies will need to address whether this is a general mechanism or whether additional pathways exist for the reestablishment of a physiological size and structure of lymphoid tissue during the contraction phase of immune responses. This is also clinically relevant as such knowledge may open venues for controlling of TLOs, which are often associated with chronic inflammation.

## References


Aasheim HC, Delabie J, Finne, EF (2005) Ephrin-A1 binding to CD4+ T lymphocytes stimulates migration and induces tyrosine phosphorylation of PYK2. Blood 105:2869–2876

Anderson AO, Shaw S (1993) T cell adhesion to endothelium: the FRC conduit system and other anatomic and molecular features which facilitate the adhesion cascade in lymph node. Semin Immunol 5:271–282

Angeli V, Ginhoux F, Llodra J, Quemeneur L, Frenette PS, Skobe M, Jessberger R, Merad M, Randolph GJ (2006) B cell-driven lymphangiogenesis in inflamed lymph nodes enhances dendritic cell mobilization. Immunity 24 203–215

Baekkevold ES, Yamanaka T, Palframan RT, Carlsen HS, Reinholt FP, von Andrian UH, Brandtzaeg P, Haraldsen G (2001) The CCR7 ligand elc (CCL19) is transcytosed in high endothelial venules and mediates T cell recruitment. J Exp Med 193:1105–1112

Bajenoff M, Egen JG, Koo LY, Laugier JP, Brau F, Glaichenhaus N, Germain RN (2006) Stromal cell networks regulate lymphocyte entry, migration, and territoriality in lymph nodes. Immunity 25:989–1001

Bargatze RF, Jutila MA, Butcher EC (1995) Distinct roles of L-selectin and integrins alpha 4 beta 7 and LFA-1 in lymphocyte homing to Peyer's patch-HEV in situ: the multistep model confirmed and refined. Immunity 3:99–108

Barreiro O, Yanez-Mo M, Serrador JM, Montoya MC, Vicente-Manzanares M, Tejedor R, Furthmayr H, Sanchez-Madrid F (2002) Dynamic interaction of VCAM-1 and ICAM-1 with moesin and ezrin in a novel endothelial docking structure for adherent leukocytes. J Cell Biol 157:1233–1245

Belisle C, Sainte-Marie G (1985) The narrowing of high endothelial venules of the rat lymph node. Anatomical Record 211:184–191

Berg EL, McEvoy LM, Berlin C, Bargatze RF, Butcher EC (1993) L-selectin-mediated lymphocyte rolling on MAdCAM-1. Nature 366:695–698

Berlin C, Berg EL, Briskin MJ, Andrew DP, Kilshaw PJ, Holzmann B, Weissman IL, Hamann A, Butcher EC (1993) Alpha 4 beta 7 integrin mediates lymphocyte binding to the mucosal vascular addressin MAdCAM-1. Cell 74:185–195. issn: 0092–8674

Berlin-Rufenach C, Otto F, Mathies M, Westermann J, Owen MJ, Hamann A, Hogg N (1999) Lymphocyte migration in lymphocyte function-associated antigen (LFA)-1-deficient mice. J Exp Med 189:1467–1478

Browning JL, Allaire N, Ngam-Ek A, Notidis E, Hunt J, Perrin S, Fava RA (2005) Lymphotoxin-beta receptor signaling is required for the homeostatic control of HEV differentiation and function. Immunity 23:539–550

Campbell FR (1983) Intercellular contacts of lymphocytes during migration across high-endothelial venules of lymph nodes. An electron microscopic study. Anat Rec 207:643–652

Carman CV, Springer TA (2008) Trans-cellular migration: cell-cell contacts get intimate. Curr Opin Cell Biol 20:533–540

Carriere V, Roussel L, Ortega N, Lacorre DA, Americh L, Aguilar L, Bouche G, Girard JP (2007) IL-33, the IL-1-like cytokine ligand for ST2 receptor, is a chromatin-associated nuclear factor in vivo. Proc Natl Acad Sci USA 104:282–287

Chtanova T, Schaeffer M, Han SJ, van Dooren GG, Nollmann M, Herzmark P, Chan SW, Satija H, Camfield K, Aaron H et al (2008) Dynamics of neutrophil migration in lymph nodes during infection. Immunity 29:487–496

Chyou S, Ekland EH, Carpenter AC, Tzeng TC, Tian S, Michaud M, Madri JA, Lu TT (2008) Fibroblast-type reticular stromal cells regulate the lymph node vasculature. J Immunol 181:3887–3896

Drayton DL, Bonizzi G, Ying X, Liao S, Karin M, Ruddle NH (2004) I kappa B kinase complex alpha kinase activity controls chemokine and high endothelial venule gene expression in lymph nodes and nasal-associated lymphoid tissue. J Immunol 173:6161–6168

Drayton DL, Liao S, Mounzer RH, Ruddle NH (2006) Lymphoid organ development: from ontogeny to neogenesis. Nat Immunol 7:344–353

Drayton DL, Ying X, Lee J, Lesslauer W, Ruddle NH (2003) Ectopic LT alpha beta directs lymphoid organ neogenesis with concomitant expression of peripheral node addressin and a HEV-restricted sulfotransferase. J Exp Med 197:1153–1163

Engelhardt B, Wolburg H (2004) Mini-review: transendothelial migration of leukocytes: through the front door or around the side of the house? Eur J Immunol 34:2955–2963

Faveeuw C, Preece G, Ager A (2001) Transendothelial migration of lymphocytes across high endothelial venules into lymph nodes is affected by metalloproteinases. Blood 98:688–695

Foster K, Sheridan J, Veiga-Fernandes H, Roderick K, Pachnis V, Adams R, Blackburn C, Kioussis D, Coles M (2008) Contribution of neural crest-derived cells in the embryonic and adult thymus. J Immunol 180:3183–3189

Galkina E, Tanousis K, Preece G, Tolaini M, Kioussis D, Florey O, Haskard DO, Tedder TF, Ager A (2003) L-selectin shedding does not regulate constitutive T cell trafficking but controls the migration pathways of antigen-activated T lymphocytes. J Exp Med 198:1323–1335

Gauguet JM, Rosen SD, Marth JD, von Andrian UH (2004) Core 2 branching beta1,6-N-acetylglucosaminyltransferase and high endothelial cell N-acetylglucosamine-6-sulfotransferase

exert differential control over B- and T-lymphocyte homing to peripheral lymph nodes. Blood 104:4104–4112
Girard JP, Springer TA (1995) Cloning from purified high endothelial venule cells of hevin, a close relative of the antiadhesive extracellular matrix protein SPARC. Immunity 2:113–123
Girard JP, Springer TA (1996) Modulation of endothelial cell adhesion by hevin, an acidic protein associated with high endothelial venules. J Biol Chem 271:4511–4517
Gretz JE, Norbury CC, Anderson AO, Proudfoot AE, Shaw S (2000) Lymph-borne chemokines and other low molecular weight molecules reach high endothelial venules via specialized conduits while a functional barrier limits access to the lymphocyte microenvironments in lymph node cortex. J Exp Med 192:1425–1440
Halin C, Tobler NE, Vigl B, Brown LF, Detmar M (2007) VEGF-A produced by chronically inflamed tissue induces lymphangiogenesis in draining lymph nodes. Blood 110:3158–3167
Harris H, Miyasaka M (1995) Reversible stimulation of lymphocyte motility by cultured high endothelial cells: mediation by L-selectin. Immunology 84:47–54
Hay JB, Hobbs BB (1977) The flow of blood to lymph nodes and its relation to lymphocyte traffic and the immune response. J Exp Med 145:31–44
Hemmerich S, Bistrup A, Singer MS, van Zante A, Lee JK, Tsay D, Peters M, Carminati JL, Brennan TJ, Carver-Moore K et al. (2001) Sulfation of L-selectin ligands by an HEV-restricted sulfotransferase regulates lymphocyte homing to lymph nodes. Immunity 15:237–247
Hendriks HR, Duijvestijn AM, Kraal G (1987) Rapid decrease in lymphocyte adherence to high endothelial venules in lymph nodes deprived of afferent lymphatic vessels. Eur J Immunol 17:1691–1695
Izawa D, Tanaka T, Saito K, Ogihara H, Usui T, Kawamoto S, Matsubara K, Okubo K, Miyasaka M (1999) Expression profile of active genes in mouse lymph node high endothelial cells. Int Immunol 11:1989–1998
Kanda H, Newton R, Klein R, Morita Y, Gunn MD, Rosen SD (2008) Autotaxin, an ectoenzyme that produces lysophosphatidic acid, promotes the entry of lymphocytes into secondary lymphoid organs. Nat Immunol 9:415–423
Kanda H, Tanaka T, Matsumoto M, Umemoto E, Ebisuno Y, Kinoshita M, Noda M, Kannagi R, Hirata T, Murai T et al. (2004) Endomucin, a sialomucin expressed in high endothelial venules, supports L-selectin-mediated rolling. Int Immunol 16:1265–1274
Kashiwazaki M, Tanaka T, Kanda H, Ebisuno Y, Izawa D, Fukuma N, Akimitsu N, Sekimizu K, Monden M, Miyasaka M (2003) A high endothelial venule-expressing promiscuous chemokine receptor DARC can bind inflammatory, but not lymphoid, chemokines and is dispensable for lymphocyte homing under physiological conditions. Int Immunol 15:1219–1227
Katakai T, Hara T, Sugai M, Gonda H, Shimizu A (2004) Lymph node fibroblastic reticular cells construct the stromal reticulum via contact with lymphocytes. J Exp Med 200:783–795
Kawashima H, Petryniak B, Hiraoka N, Mitoma J, Huckaby V, Nakayama J, Uchimura K, Kadomatsu K, Muramatsu T, Lowe JB, Fukuda M (2005) N-acetylglucosamine-6-O-sulfotransferases 1 and 2 cooperatively control lymphocyte homing through L-selectin ligand biosynthesis in high endothelial venules. Nat Immunol 6:1096–1104
Klinger A, Gebert A, Bieber K, Kalies K, Ager A, Bell EB, Westermann J (2009) Cyclical expression of L-selectin (CD62L) by recirculating T cells. Int Immunol 21:443–455
Kumar V, Scandella E, Danuser R, Onder L, Fukui Y, Ludewig B, Stein, JV (2010) Mesoscopic analysis of B cell/lymphotoxin-dependent global lymphoid tissue remodeling during viral infection. Blood 115(23):4725–33. Epub 2010 Feb 25
Lacorre DA, Baekkevold ES, Garrido I, Brandtzaeg P, Haraldsen G, Amalric F, Girard JP (2004) Plasticity of endothelial cells: rapid dedifferentiation of freshly isolated high endothelial venule endothelial cells outside the lymphoid tissue microenvironment. Blood 103:4164–4172
Lammert E, Cleaver O, Melton D (2001) Induction of pancreatic differentiation by signals from blood vessels. Science 294:564–567

Lehmann JC, Jablonski-Westrich D, Haubold U, Gutierrez-Ramos JC, Springer T, Hamann A (2003) Overlapping and selective roles of endothelial intercellular adhesion molecule-1 (ICAM-1) and ICAM-2 in lymphocyte trafficking. J Immunol 171:2588–2593

Liao S, Bentley K, Lebrun M, Lesslauer W, Ruddle FH, Ruddle NH (2007) Transgenic LacZ under control of Hec-6st regulatory sequences recapitulates endogenous gene expression on high endothelial venules. Proc Natl Acad Sci USA 104:4577–4582

Liao S, Ruddle NH (2006) Synchrony of high endothelial venules and lymphatic vessels revealed by immunization. J Immunol 177:3369–3379

M'Rini C, Cheng G, Schweitzer C, Cavanagh LL, Palframan RT, Mempel TR, Warnock RA, Lowe JB, Quackenbush EJ, von Andrian UH (2003) A novel endothelial L-selectin ligand activity in lymph node medulla that is regulated by alpha(1,3)-fucosyltransferase-IV. J Exp Med 198:1301–1312

Marchesi VT, Gowans JL (1964) The migration of lymphocytes through the endothelium of venules in lymph nodes: an electron microscope study. Proc Roy Soc Ser B 159:283–290

Martin-Fontecha A, Thomsen LL, Brett S, Gerard C, Lipp M, Lanzavecchia A, Sallusto F (2004) Induced recruitment of NK cells to lymph nodes provides IFN-gamma for T(H)1 priming. Nat Immunol 5:1260–1265

Mebius RE (2003) Organogenesis of lymphoid tissues. Nat Rev Immunol 3:292–303

Mebius RE, Dowbenko D, Williams A, Fennie C, Lasky LA, Watson SR (1993) Expression of GlyCAM-1, an endothelial ligand for L-selectin, is affected by afferent lymphatic flow. J Immunol 151:6769–6776

Mebius RE, Streeter PR, Breve J, Duijvestijn AM, Kraal G (1991) The influence of afferent lymphatic vessel interruption on vascular addressin expression. J Cell Biol 115:85–95

Mebius RE, Streeter PR, Michie S, Butcher EC, Weissman IL (1996) A developmental switch in lymphocyte homing receptor and endothelial vascular addressin expression regulates lymphocyte homing and permits CD4+ CD3- cells to colonize lymph nodes. Proc Natl Acad Sci USA 93:11019–11024

Mebius RE, Watson SR (1993) L- and E-selectin can recognize the same naturally occurring ligands on high endothelial venules. J Immunol 151:3252–3260

Mempel TR, Henrickson SE, Von Andrian UH (2004) T-cell priming by dendritic cells in lymph nodes occurs in three distinct phases. Nature 427:154–159

Mitoma J, Bao X, Petryanik B, Schaerli P, Gauguet JM, Yu SY, Kawashima H, Saito H, Ohtsubo K, Marth JD et al. (2007) Critical functions of N-glycans in L-selectin-mediated lymphocyte homing and recruitment. Nat Immunol 8:409–418

Miyasaka M, Tanaka T (2004) Lymphocyte trafficking across high endothelial venules: dogmas and enigmas. Nat Rev Immunol 4:360–370

Mueller SN, Hosiawa-Meagher KA, Konieczny BT, Sullivan BM, Bachmann, MF, Locksley RM, Ahmed R, Matloubian M (2007) Regulation of homeostatic chemokine expression and cell trafficking during immune responses. Science 317:670–674

Nakano H, Gunn MD (2001) Gene duplications at the chemokine locus on mouse chromosome 4: multiple strain-specific haplotypes and the deletion of secondary lymphoid-organ chemokine and EBI-1 ligand chemokine genes in the plt mutation. J Immunol 166:361–369

Nakasaki T, Tanaka T, Okudaira S, Hirosawa M, Umemoto E, Otani K, Jin S, Bai Z, Hayasaka H, Fukui Y et al (2008) Involvement of the lysophosphatidic acid-generating enzyme autotaxin in lymphocyte-endothelial cell interactions. Am J Pathol 173:1566–1576

Okada T, Ngo VN, Ekland EH, Forster R, Lipp M, Littman DR, Cyster JG (2002) Chemokine requirements for B cell entry to lymph nodes and Peyer's patches. J Exp Med 196:65–75

Palframan RT, Jung S, Cheng G, Weninger W, Luo Y, Dorf M, Littman DR, Rollins BJ, Zweerink H, Rot A, von Andrian UH (2001) Inflammatory chemokine transport and presentation in HEV: a remote control mechanism for monocyte recruitment to lymph nodes in inflamed tissues. J Exp Med 194:1361–1373

Pfeiffer F, Kumar V, Butz S, Vestweber D, Imhof BA, Stein JV, Engelhardt B (2008) Distinct molecular composition of blood and lymphatic vascular endothelial cell junctions establishes specific functional barriers within the peripheral lymph node. Eur J Immunol 38:2142–2155

Randolph GJ, Angeli V, Swartz MA (2005) Dendritic-cell trafficking to lymph nodes through lymphatic vessels. Nat Rev Immunol 5:617–628

Rosen SD (2004) Ligands for L-selectin: homing, inflammation, and beyond. Annu Rev Immunol 22:129–156

Sabin F (1913) The origin and development of the lymphatic system. Hospital reports monograph new series no. V. The John Hopkins Press, pp 65–70

Scandella E, Bolinger B, Lattmann E, Miller S, Favre S, Littman DR, Finke D, Luther SA, Junt T, Ludewig B (2008) Restoration of lymphoid organ integrity through the interaction of lymphoid tissue-inducer cells with stroma of the T cell zone. Nat Immunol 9:667–675

Scimone ML, Felbinger TW, Mazo IB, Stein JV, Von Andrian UH, Weninger W (2004) CXCL12 mediates CCR7-independent homing of central memory cells, but not naive T cells, in peripheral lymph nodes. J Exp Med 199:1113–1120

Sixt M, Kanazawa N, Selg M, Samson T, Roos G, Reinhardt DP, Pabst R, Lutz MB, Sorokin L (2005) The conduit system transports soluble antigens from the afferent lymph to resident dendritic cells in the T cell area of the lymph node. Immunity 22:19–29

Smith ME, Ford WL (1983) The recirculating lymphocyte pool of the rat: a systematic description of the migratory behaviour of recirculating lymphocytes. Immunology 49:83–94

Soderberg KA, Payne GW, Sato A, Medzhitov R, Segal SS, Iwasaki A (2005) Innate control of adaptive immunity via remodeling of lymph node feed arteriole. Proc Natl Acad Sci USA 102:16315–16320

Stein JV, Rot A, Luo Y, Narasimhaswamy M, Nakano H, Gunn MD, Matsuzawa A, Quackenbush EJ, Dorf ME, von Andrian UH (2000) The CC chemokine thymus-derived chemotactic agent 4 (TCA-4, secondary lymphoid tissue chemokine, 6Ckine, exodus-2) triggers lymphocyte function-associated antigen 1-mediated arrest of rolling T lymphocytes in peripheral lymph node high endothelial venules. J Exp Med 191:61–76

Streeter PR, Rouse BT, Butcher EC (1988) Immunohistologic and functional characterization of a vascular addressin involved in lymphocyte homing into peripheral lymph nodes. J Cell Biol 107:1853–1862

Swarte VV, Joziasse DH, Van den Eijnden DH, Petryniak B, Lowe JB, Kraal G, Mebius RE (1998) Regulation of fucosyltransferase-VII expression in peripheral lymph node high endothelial venules. Eur J Immunol 28:3040–3047

Tang ML, Steeber DA, Zhang XQ, Tedder TF (1998) Intrinsic differences in L-selectin expression levels affect T and B lymphocyte subset-specific recirculation pathways. J Immunol 160:5113–5121

Uchimura K, Gauguet JM, Singer MS, Tsay D, Kannagi R, Muramatsu T, von Andrian UH, Rosen SD (2005) A major class of L-selectin ligands is eliminated in mice deficient in two sulfotransferases expressed in high endothelial venules. Nat Immunol 6:1105–1113

Umemoto E, Tanaka T, Kanda H, Jin S, Tohya K, Otani K, Matsutani T, Matsumoto M, Ebisuno Y, Jang MH et al (2006) Nepmucin, a novel HEV sialomucin, mediates L-selectin-dependent lymphocyte rolling and promotes lymphocyte adhesion under flow. J Exp Med 203:1603–1614

van Tuyl M, Groenman F, Wang J, Kuliszewski M, Liu J, Tibboel D, Post M (2007) Angiogenic factors stimulate tubular branching morphogenesis of sonic hedgehog-deficient lungs. Dev Biol 303:514–526

van Zante A, Gauguet JM, Bistrup A, Tsay D, von Andrian UH, Rosen SD (2003) Lymphocyte-HEV interactions in lymph nodes of a sulfotransferase-deficient mouse. J Exp Med 198:1289–1300

Vassileva G, Soto H, Zlotnik A, Nakano H, Kakiuchi T, Hedrick JA, Lira SA (1999) The reduced expression of 6Ckine in the plt mouse results from the deletion of one of two 6Ckine genes. J Exp Med 190:1183–1188

Vestweber D (2007) Adhesion and signaling molecules controlling the transmigration of leukocytes through endothelium. Immunol Rev 218:178–196 von Andrian UH, M'Rini C (1998) In situ analysis of lymphocyte migration to lymph nodes. Cell Adhes Commun 6:85–96

Warnock RA, Askari S, Butcher EC, von Andrian UH (1998) Molecular mechanisms of lymphocyte homing to peripheral lymph nodes. J Exp Med 187:205–216

Warnock RA, Campbell JJ, Dorf ME, Matsuzawa A, McEvoy LM, Butcher EC (2000) The role of chemokines in the microenvironmental control of T versus B cell arrest in Peyer's patch high endothelial venules. J Exp Med 191:77–88

Webster B, Ekland EH, Agle LM, Chyou S, Ruggieri R, Lu TT (2006) Regulation of lymph node vascular growth by dendritic cells. J Exp Med 203:1903–1913

# Part III
# Programmed and Nascent Gut-Associated Organized Lymphoid Tissues

# Chapter 9
# Structure and Development of Peyer's Patches in Humans and Mice

**Tom Cupedo, Mark C. Coles, and Henrique Veiga-Fernandes**

**Abstract** Peyer's patches are lymphoid organs situated on the anti-mesenteric side of the mid-intestine. Within the Peyer's patches, immune responses to intestinal-derived antigens are initiated. In this chapter, we will discuss the structure and function of the Peyer's patches. In addition, the development of Peyer's patches during fetal life will be reviewed, with an emphasis on the reciprocal interaction between distinct hematopoietic cell subsets and their stromal environment in the Peyer's patch anlage.

## 9.1 Structure of the Peyer's Patches

The intestinal immune system harbours two types of lymphoid tissues (1) programmed lymphoid tissues that form during fetal development and (2) induced lymphoid tissues that assemble after birth and are directed by environmental cues. The programmed tissues are the mesenteric lymph nodes and the Peyer's patches, while the cryptopatches and isolated lymphoid follicles (ILF) make up the inducible arm of the intestinal immune tissues. Here we will focus on the programmed tissues and in particular the Peyer's patches.

Peyer's patches are specialised secondary lymphoid organs located in the anti-mesenteric side of the mid-intestine. In mice, these structures form in variable numbers ranging from 5 to 12 (Nishikawa et al. 2003). Peyer's patch primordia start to develop in the duodenum and proximal ileum with the most distal anlagen

T. Cupedo
Department of Hematology, Erasmus University Medical Center, Rotterdam, The Netherlands

M.C. Coles
Centre for Immunology and Infection, Department of Biology and Hull York Medical School, University of York, York, UK

H. Veiga-Fernandes (✉)
Immunobiology Unit, Instituto de Medicina Molecular, Faculdade de Medicina de Lisboa, Lisboa, Portugal
e-mail: jhfernandes@fm.ul.pt

P. Balogh (ed.), *Developmental Biology of Peripheral Lymphoid Organs*,
DOI 10.1007/978-3-642-14429-5_9, 

structures being formed subsequently at relatively regular intervals in the lower mid-gut (Adachi et al. 1997).

The structure of adult Peyer's patches is similar to that of lymph nodes, with B and T cells segregating in distinct organised areas (Griebel and Hein 1996). In both lymph nodes and Peyer's patches, this level of organisation occurs in total absence of infection since germfree mice also display this pattern of segregation (Crabbe et al. 1970). However, the organisation of mature Peyer's patches differs from lymph nodes in other respects. As an example, the luminal side of the Peyer's patches is represented by a protrusion covered by a layer of follicle-associated epithelium that includes specialised epithelial cells, designated M cells, which also mediate immune functions (Corr et al. 2008). Additionally, Peyer's patches lack afferent lymphatics, but have efferent lymphatics that originate in lymphatic sinuses on the serosal side of the Peyer's patches, draining the lymph and immune cells to the mesenteric lymph node and then to the thoracic duct (Pellas and Weiss 1990).

As for other secondary lymphoid organs, HEVs are important structures in Peyer's patches, allowing for lymphocyte homing. A key molecule in this process is MAdCAM-1, which is expressed also at early stages of lymph node development (Mebius et al. 1996). Detection of MAdCAM-1 in the developing Peyer's patch primordium starts at E16.5 and at E17.5, a reticulated pattern of expression emerges (Hashi et al. 2001). Just after birth, MAdCAM-1 expression finally surrounds and associates with the follicle structures formed by IL7R$\alpha$, CD11c and VCAM-1 positive cells (Hashi et al. 2001). Interestingly, the appearance of the first mature lymphocytes in the Peyer's patch anlagen occurs prior to the establishment of this MAdCAM-1 organised network at E18.5. B and T cells are initially distributed homogeneously, but B cells start to aggregate into follicles just after birth assuming the definitive distribution pattern seen in mature Peyer's patches (Hashi et al. 2001).

The luminal side of Peyer's patches is covered by the follicle-associated epithelium, which is characterised by a high density of M cells, forming the interface between the enteric lymphoid organs and the luminal environment. While the apical surface of M cells has a poorly organised brush border (Kerneis et al. 1997), M cells possess unique intra-epithelial invaginations at their basolateral borders, which contain macrophages, dendritic cells and B and T lymphocytes (Neutra et al. 1996); hence these "pockets" facilitate the contact between luminal antigens and Peyer's patch immune cells. Apart from Peyer's patches, M cells are also found in enteric ILF, the appendix and in other non-enteric mucosal-associated lymphoid tissues (Corr et al. 2008).

The primary function of M cells is trans-epithelial transport of diverse substances (Gebert et al. 1996), but they have also been involved in transport of microorganisms in a receptor-mediated process (Fotopoulos et al. 2002; Tyrer et al. 2007). There is also evidence that M cells may help promoting immune responses. In fact, M cells can produce IL-1 upon LPS stimulation, which in turn induces T cell proliferation (Pappo and Mahlman 1993). M cells also express pattern recognition receptors, such as TLR4 (Tyrer et al. 2007) and triggering of these receptors increase particle uptake and DC recruitment (Chabot et al. 2006).

Additionally, secretory IgA has a high affinity to M cells (Mantis et al. 2002), which in turn can translocate these complexes to the sub-epithelial areas (Kadaoui and Corthesy 2007). Thus, this unique mucosal cell type act has a frontline for Ag sampling and up-taking for the enteric immune system, transporting Ag to the underlying Peyer's patch tissue but also mediating and contributing to the initiation of immune responses.

Peyer's patches are also the residence for a large number of Immunoglobin A (IgA) B cells. IgA is the most abundant immunoglobin isotype produced in the body, and it was calculated that 80% of all IgA secreting B cells reside in the gut mucosa (Macpherson and Slack 2007; Suzuki et al. 2007). Once generated in the Peyer's patches, IgA B cells migrate to the draining mesenteric lymph node where they further differentiate into IgA secreting cells, which in turn migrate to the enteric lamina propria (Cyster 2003). Besides their role in neutralising toxins and pathogens (Lycke et al. 1999; Wijburg et al. 2006), IgA may also help preventing local inflammation and immune responses to commensal flora or non-pathogenic agents (Macpherson and Uhr 2004; Brandtzaeg et al. 2006; Macpherson and Slack 2007).

A significant part of IgA B cells is generated in a T cell-dependent manner in the Peyer's patch germinal centres (GC) and this process partly depends on follicular CD4 helper T cells. In fact, key markers of Ig class switch, such as AID (Muramatsu et al. 2000) and germline alpha transcripts (Kinoshita et al. 2001), are highly expressed in Peyer's patches when compared with mesenteric lymph nodes or the peritoneal cavity (Bergqvist et al. 2006). Interestingly, the Peyer's patch environment, rich in IL6, IL21 and activated B cells (Tsuji et al. 2009), was also shown to preferentially support the differentiation of suppressive Foxp3$^{+}$ T cells into follicular helper T cells (Tsuji et al. 2009). Furthermore, locally produced TGF-β, IL4, IL6 and IL10 can help expanding IgA secreting cells (Cerutti and Rescigno 2008).

## 9.2 Hematopoietic Cells in Peyer's Patch Development

As for lymph node development, the cellular mechanisms implicated in Peyer's patch development are well characterised, relying on interactions between cells from hematopoietic and mesenchymal origin. Despite the parallels between lymph node and Peyer's patch genesis, these processes are not entirely identical and even require differential players. In mice, although the first B cell follicles are detected after birth (Villena et al. 1983; Yoshida et al. 1995), the initial set-up of Peyer's patch architecture starts during embryonic life (Hashi et al. 2001). Hematopoietic cells that initially colonise the intestine include CD4$^{+}$CD3$^{-}$IL7R$\alpha^{+}$c-Kit$^{+}$ LTi cells (Adachi et al. 1997, 1998; Yoshida et al. 1999) and a phenotypically distinct population of CD4$^{-}$CD3$^{-}$c-Kit$^{+}$IL7R$\alpha^{-}$CD11c$^{+}$ Lymphoid Tissue initiator (LTin) cells (Fukuyama and Kiyono 2007; Veiga-Fernandes et al. 2007). Both LTi and LTin aggregate together with mesenchymal origin VCAM-1$^{+}$ICAM-1$^{+}$ LTo cells (Hashi et al. 2001; Honda et al. 2001) forming the Peyer's patch primordium

(Adachi et al. 1997; Yoshida et al. 1999; Hashi et al. 2001; Veiga-Fernandes et al. 2007). While lymph node localisation is seemingly determined before the migration of haematopoietic cells to prospective sites of development, location of Peyer's patch anlagen does not seem to be strictly pre-determined.

By E15.5, increasing numbers of highly motile hematopoietic cells are found evenly distributed throughout the gut, and by E16.5, these cells aggregate to form the Peyer's patch primordial (Adachi et al. 1997, 1998; Yoshida et al. 1999; Veiga-Fernandes et al. 2007). LTi cells and VCAM-1$^+$ LTo cells initially distribute diffusely throughout the primordium at E17.5, while CD11c$^+$ cells appear at the periphery of the Peyer's patch primordium (Hashi et al. 2001). One day later, at E18.5, IL7R$\alpha$ CD11c$^+$ and VCAM-1$^+$ cells form irregular segregated regions that, around birth, eventually form discreet sub-regions of similar size. However, in these sub-regions, each cell type displays a different distribution, with IL7R$\alpha^+$ cells occupying the centre, while VCAM-1$^+$ and CD11c$^+$ cells preferentially accumulate in the periphery (Hashi et al. 2001). Since IL7R$\alpha^+$ and CD11c$^+$ cells display a different pattern of the chemokine receptors CXCR5 and CCR7 (Hashi et al. 2001), it is possible that these cells may have different affinities for areas with different expression for the respective chemokines, but further analyses are required to elucidate this point. Thus, although the relationship between CD11c$^+$ cells observed at late stages of development (Adachi et al. 1997; Yoshida et al. 1999; Hashi et al. 2001) and E15.5 early CD11c$^+$ LTin cells is unclear (Fukuyama and Kiyono 2007; Veiga-Fernandes et al. 2007), it is likely that the three main cellular partners of Peyer's patch induction, LTi, LTin and LTo cells, are sufficient to drive the initial architecture of Peyer's patches. Supporting this idea is the fact that the initial process of Peyer's patch organisation is totally independent of B or T cells, since in *scid/scid* mice, the initial architecture set-up occurs normally (Hashi et al. 2001).

Haematopoietic LTi and LTin cells play important roles in Peyer's patch development. Hence, in the absence of LTi cells, which occur in *Id2*- (Yokota et al. 1999) and *Ror$\gamma$*- (Sun et al. 2000) deficient mice, Peyer's patches fail to develop. Moreover, adoptive transfer of LTi cells into neonatal mice with minute numbers of Peyer's patches was shown to rescue the organogenesis of these structures (Finke et al. 2002). In agreement with this idea, increased LTi cell numbers, obtained by IL-7 over expression, result in high number of Peyer's patches (Meier et al. 2007). Concerning LTin cells, selective and partial ablation of CD11c$^+$ cells results in impaired Peyer's patch development, and mice deficient for the tyrosine kinase receptor RET, expressed by LTin cells, do not form Peyer's patches (Veiga-Fernandes et al. 2007). Most importantly, the use of the RET ligand Artemin (ARTN) in explants organ cultures of embryonic intestines induced LTin and LTi clustering and upregulation of VCAM-1 by mesenchymal cells, resulting in ectopic Peyer's patch primordia (Veiga-Fernandes et al. 2007). During lymph node development, CD11c$^+$ cells were also described in the anlagen (Yoshida et al. 2002); however, the relationship between lymph node and enteric CD11c$^+$ cells is unclear and it is not known what, if any, functions these cells may play in the context of lymph node genesis.

While most molecules implicated in lymph node development also play a role in Peyer's patch formation, some signalling pathways are not equally used in both processes. As an example, while IL7R signal is crucial for Peyer's patch development, as revealed by *Il7r*$^{-/-}$ mice (Adachi et al. 1998; Yoshida et al. 1999), brachial, axillary and mesenteric lymph node develop normally in these animals (Mebius 2003). Similarly, while in the absence of RET signalling Peyer's patches fail to form, lymph node development in *Ret*$^{-/-}$ mice is seemingly normal (Veiga-Fernandes et al. 2007). On the other hand, the differential use of molecular pathways is also revealed by mutants of the RANKL-RANK signalling axis. While in *rankl*$^{-/-}$ (Dougall et al. 1999; Kong et al. 1999; Kim et al. 2000) and *Traf6*$^{-/-}$ (Naito et al. 1999) mice, lymph node development is severely compromised, Peyer's patch development is entirely normal. Hence, although lymph node and Peyer's patch genesis display obvious parallels, the differential requirements for their development may also reflect distinct genetic signatures of their respective cellular players (Fig. 9.1).

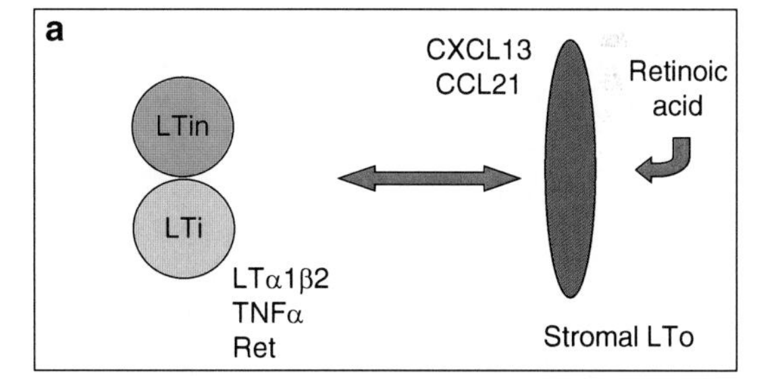


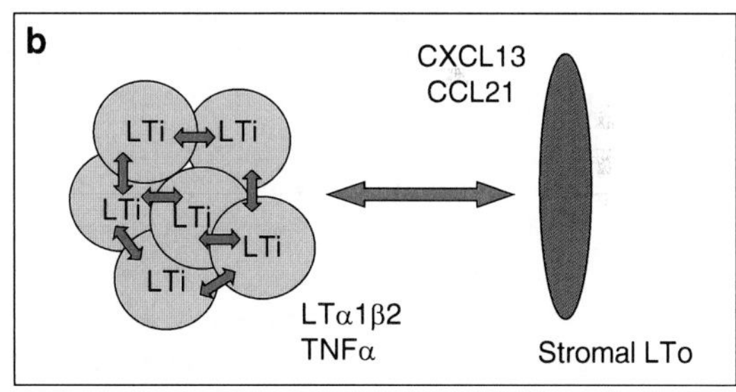


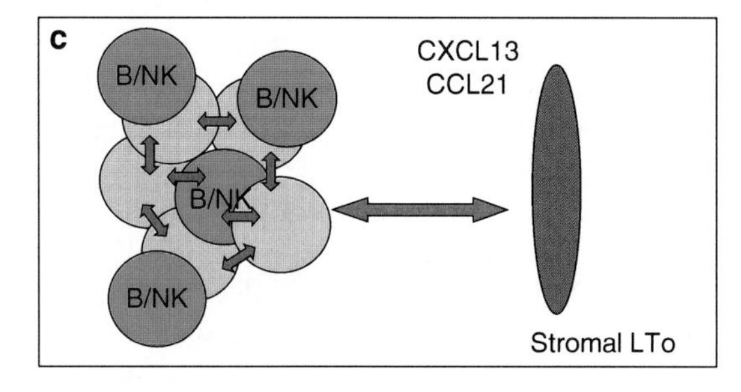


**Fig. 9.1** Cellular interactions during Peyer's Patch development. (**a**) Stromal cells at future Peyer's patch locations secrete CXCL13, CCL21, and a ligand for the tyrosine kinase receptor RET. What induces the localised production of these chemotactic proteins is currently unknown. LTi cells and RET-expressing CD11c+ LTin cells are attracted to the Peyer's patch primordium and will activate stromal cells via LTα1β2–LTβR and TNFα–TNFR interactions. (**b**) Continued production of CXCL13 and CCL21 will attract additional circulating LTi cells and their presumptive precursors. These will engage in paracrine interactions as well as in interactions with the stroma. (**c**) In the early postnatal stages, T and B cells will flood the Peyer's patch and start to contribute to the stromal crosstalk in order to perfect the Peyer's patch architecture

## 9.3 Stromal Cells in Peyer's Patch Development

Similar to the development of lymph nodes, productive interactions between haematopoietic and mesenchymal cells are also at the centre of normal Peyer's patch development. VCAM-1$^{+}$ICAM-1$^{+}$ LTo cells aggregate together with LTi and LTin cells, so that discreet VCAM-1$^{+}$ patches are unequivocally found in the intestine from E16.5 onwards (Adachi et al. 1997; Yoshida et al. 1999; Hashi et al. 2001; Veiga-Fernandes et al. 2007). Although rare VCAM-1$^{+}$ aggregates could be identified at E15.5, prior to haematopoietic cell accumulation (Adachi et al. 1997), VCAM-1 expression levels at this stage were negligible when compared with E17.5 (Adachi et al. 1997), strongly indicating that full differentiation of mesenchymal cells requires further interaction with haematopoietic cells. In early phases of Peyer's patch development, LTo, LTi and LTin cells are evenly distributed in the primordium (Hashi et al. 2001; Veiga-Fernandes et al. 2007); however, LTo cells are likely to also mediate Peyer's patch organisation since they are among the first cells to segregate into follicle-like structures around E18.5 (Hashi et al. 2001). As for Peyer's patch haematopoietic cells, increasing evidence suggests that LTo cells are very heterogeneous. In fact, as described in developing lymph nodes (Cupedo et al. 2004), Peyer's patch stromal cells can be subdivided in VCAM-1$^{hi}$ICAM-1$^{hi}$ and VCAM-1$^{med}$ICAM-1$^{med}$ (Okuda et al. 2007). However, it remains to be determined if similarly to lymph node LTo cells, Peyer's patch VCAM-1$^{hi}$ICAM-1$^{hi}$ cells express higher levels of the homeostatic chemokines CXCL13, CCL21 and CCL19, when compared with VCAM-1$^{med}$ICAM-1$^{med}$ cells (Cupedo et al. 2004). Nevertheless, LTo heterogeneity within the Peyer's patches suggests that different LTo cells might provide different cues to resident haematopoietic cells. Despite their similarities, LTo cells from Peyer's patches and mesenteric lymph nodes also exhibit distinct characteristics. While LTo cells from mesenteric lymph nodes have surface expression of TRANCE (Cupedo et al. 2004; Okuda et al. 2007), their Peyer's patch counterparts lack surface expression of this ligand (Okuda et al. 2007). Using microarray analysis, Okuda et al. further compared LTo cells from mesenteric lymph nodes and Peyer's patches revealing that their respective genetic signatures are distinct. As an example, mesenteric lymph node LTo cells express significantly higher levels of cytokines and chemokines such as IL6, IL7, CCL7, CXCL1 and CCL11(Okuda et al. 2007). On the other hand, the homeostatic chemokines CCL21, CCL19 and CXCL13 were more abundant in Peyer's patch LTo cells (Okuda et al. 2007). Compared with their Peyer's patch counterparts, mesenteric lymph node LTo cells also display higher levels of transcription factors such as Meox2, Lhx8 and Prrx1 (Okuda et al. 2007), but their functional significance in lymph node development remains elusive. These distinct gene expression profiles are consistent with the notion that the differential activity of organiser cells may determine the use of different molecular axis in different organs, such as the TRANCE and IL7R signalling in lymph nodes and Peyer's patches, respectively. Nevertheless, it remains to be determined if these different expression profiles are cell autonomous or induced by different cellular interactions and to which extent the difference in developmental stages at which lymph nodes and Peyer's patches were analysed influence the observed regional differences.

The network of stromal cells persists throughout adulthood in mature Peyer's patches. Similarly to lymph nodes, follicular dendritic cells (FDCs) of mesenchymal origin are detected around the periphery and within the B cell areas of Peyer's patches (Finke 2009). In the lymph node, fibroblast reticular cells (FRCs) locate in T cell areas (Katakai et al. 2004a, b), produce IL-7 (Link et al. 2007) and create the conduit system (Gretz et al. 1996; Sixt et al. 2005). It is currently unknown if conduits exist in Peyer's patches and whether they mediate lymphocyte trafficking and Ag transport in enteric lymphoid organs. Marginal reticular cells (MRCs) (Katakai et al. 2008), of mesenchymal origin, have also been described in the sub-epithelial dome of Peyer's patches, but it is unclear if they contribute to maintenance or recruitment of immune cells upon Ag up-take in the Peyer's patch dome. IgA class switch recombination also occurs in enteric ILF (Cerutti and Rescigno 2008). However, mice treated in utero with both TNFR55-Ig and LTβR-Ig, which lack Peyer's patches and mesenteric lymph nodes but retain normal ILFs, fail to induce Ag-specific IgA responses after oral immunisation despite normal IgA enteric levels (Yamamoto et al. 2004). Thus, Peyer's patches are major sites for the induction of IgA antigen-specific responses.

## References

Adachi S, Yoshida H, Honda K, Maki K, Saijo K, Ikuta K, Saito T, Nishikawa SI (1998) Essential role of IL-7 receptor alpha in the formation of Peyer's patch anlage. Int Immunol 10:1–6

Adachi S, Yoshida H, Kataoka H, Nishikawa S (1997) Three distinctive steps in Peyer's patch formation of murine embryo. Int Immunol 9:507–514

Bergqvist P, Gardby E, Stensson A, Bemark M, Lycke NY (2006) Gut IgA class switch recombination in the absence of CD40 does not occur in the lamina propria and is independent of germinal centers. J Immunol 177:7772–7783

Brandtzaeg P, Carlsen HS, Halstensen TS (2006) The B-cell system in inflammatory bowel disease. Adv Exp Med Biol 579:149–167

Cerutti A, Rescigno M (2008) The biology of intestinal immunoglobulin A responses. Immunity 28:740–750

Chabot S, Wagner JS, Farrant S, Neutra MR (2006) TLRs regulate the gatekeeping functions of the intestinal follicle-associated epithelium. J Immunol 176:4275–4283

Corr SC, Gahan CC, Hill C (2008) M-cells: origin, morphology and role in mucosal immunity and microbial pathogenesis. FEMS Immunol Med Microbiol 52:2–12

Crabbe PA, Nash DR, Bazin H, Eyssen H, Heremans JF (1970) Immunohistochemical observations on lymphoid tissues from conventional and germ-free mice. Lab Invest 22:448–457

Cupedo T, Vondenhoff MF, Heeregrave EJ, De Weerd AE, Jansen W, Jackson DG, Kraal G, Mebius RE (2004) Presumptive lymph node organizers are differentially represented in developing mesenteric and peripheral nodes. J Immunol 173:2968–2975

Cyster JG (2003) Homing of antibody secreting cells. Immunol Rev 194:48–60

Dougall WC, Glaccum M, Charrier K, Rohrbach K, Brasel K, De Smedt T, Daro E, Smith J, Tometsko ME, Maliszewski CR, Armstrong A, Shen V, Bain S, Cosman D, Anderson D, Morrissey PJ, Peschon JJ, Schuh J (1999) RANK is essential for osteoclast and lymph node development. Genes Dev 13:2412–2424

Finke D (2009) Induction of intestinal lymphoid tissue formation by intrinsic and extrinsic signals. Semin Immunopathol 31:151–169

Finke D, Acha-Orbea H, Mattis A, Lipp M, Kraehenbuhl J (2002) $CD4^{+}CD3^{-}$ cells induce Peyer's patch development. Role of a4b1 integrin activation by CXCR5. Immunity 17:363

Fotopoulos G, Harari A, Michetti P, Trono D, Pantaleo G, Kraehenbuhl JP (2002) Transepithelial transport of HIV-1 by M cells is receptor-mediated. Proc Natl Acad Sci USA 99:9410–9414

Fukuyama S, Kiyono H (2007) Neuroregulator RET initiates Peyer's-patch tissue genesis. Immunity 26:393–395

Gebert A, Rothkotter HJ, Pabst R (1996) M cells in Peyer's patches of the intestine. Int Rev Cytol 167:91–159

Gretz JE, Kaldjian EP, Anderson AO, Shaw S (1996) Commentary: Sophisticated strategies for information encounter in the lymph node: the reticular network as a conduit of soluble information and a highway for cell traffic. J Immunol 157:495–499

Griebel PJ, Hein WR (1996) Expanding the role of Peyer's patches in B-cell ontogeny. Immunol Today 17:30–39

Hashi H, Yoshida H, Honda K, Fraser S, Kubo H, Awane M, Takabayashi A, Nakano H, Yamaoka Y, Nishikawa SI (2001) Compartmentalization of Peyer's patch anlagen before lymphocyte entry. J Immunol 166:3702–3709

Kadaoui KA, Corthesy B (2007) Secretory IgA mediates bacterial translocation to dendritic cells in mouse Peyer's patches with restriction to mucosal compartment. J Immunol 179:7751–7757

Katakai T, Hara T, Lee J-H, Gonda H, Sugai M, Shimizu A (2004a) A novel reticular stromal structure in lymph node cortex: an immuno-platform for interactions among dendritic cells, T cells and B cells. Int Immunol 16:1133–1142

Katakai T, Hara T, Sugai M, Gonda H, Shimizu A (2004b) Lymph node fibroblastic reticular cells construct the stromal reticulum via contact with lymphocytes. J Exp Med 200:783–795

Katakai T, Suto H, Sugai M, Gonda H, Togawa A, Suematsu S, Ebisuno Y, Katagiri K, Kinashi T, Shimizu A (2008) Organizer-like reticular stromal cell layer common to adult secondary lymphoid organs. J Immunol 181:6189–6200

Kerneis S, Bogdanova A, Kraehenbuhl JP, Pringault E (1997) Conversion by Peyer's patch lymphocytes of human enterocytes into M cells that transport bacteria. Science 277:949–952

Kim D, Mebius RE, MacMicking JD, Jung S, Cupedo T, Castellanos Y, Rho J, Wong BR, Josien R, Kim N, Rennert PD, Choi Y (2000) Regulation of peripheral lymph node genesis by the tumor necrosis factor family member TRANCE. J Exp Med 192:1467–1478

Kinoshita K, Harigai M, Fagarasan S, Muramatsu M, Honjo T (2001) A hallmark of active class switch recombination: transcripts directed by I promoters on looped-out circular DNAs. Proc Natl Acad Sci USA 98:12620–12623

Kong YY, Yoshida H, Sarosi I, Tan HL, Timms E, Capparelli C, Morony S, Oliveira-dos-Santos AJ, Van G, Itie A, Khoo W, Wakeham A, Dunstan CR, Lacey DL, Mak TW, Boyle WJ, Penninger JM (1999) OPGL is a key regulator of osteoclastogenesis, lymphocyte development and lymph-node organogenesis. Nature 397:315–323

Link A, Vogt TK, Favre S, Britschgi MR, Acha-Orbea H, Hinz B, Cyster JG, Luther SA (2007) Fibroblastic reticular cells in lymph nodes regulate the homeostasis of naive T cells. Nat Immunol 8:1255–1265

Lycke N, Erlandsson L, Ekman L, Schon K, Leanderson T (1999) Lack of J chain inhibits the transport of gut IgA and abrogates the development of intestinal antitoxic protection. J Immunol 163:913–919

Macpherson AJ, Slack E (2007) The functional interactions of commensal bacteria with intestinal secretory IgA. Curr Opin Gastroenterol 23:673–678

Macpherson AJ, Uhr T (2004) Induction of protective IgA by intestinal dendritic cells carrying commensal bacteria. Science 303:1662–1665

Mantis NJ, Cheung MC, Chintalacharuvu KR, Rey J, Corthesy B, Neutra MR (2002) Selective adherence of IgA to murine Peyer's patch M cells: evidence for a novel IgA receptor. J Immunol 169:1844–1851

Mebius RE (2003) Organogenesis of lymphoid tissues. Nat Rev Immunol 3:292–303
Mebius RE, Streeter PR, Michie S, Butcher EC, Weissman IL (1996) A developmental switch in lymphocyte homing receptor and endothelial vascular addressin expression regulates lymphocyte homing and permits CD4+ CD3− cells to colonize lymph nodes. Proc Natl Acad Sci USA 93:11019–11024
Meier D, Bornmann C, Chappaz S, Schmutz S, Otten LA, Ceredig R, Acha-Orbea H, Finke D (2007) Ectopic lymphoid-organ development occurs through interleukin 7-mediated enhanced survival of lymphoid-tissue-inducer cells. Immunity 26:643–654
Muramatsu M, Kinoshita K, Fagarasan S, Yamada S, Shinkai Y, Honjo T (2000) Class switch recombination and hypermutation require activation-induced cytidine deaminase (AID), a potential RNA editing enzyme. Cell 102:553–563
Naito A, Azuma S, Tanaka S, Miyazaki T, Takaki S, Takatsu K, Nakao K, Nakamura K, Katsuki M, Yamamoto T, Inoue J (1999) Severe osteopetrosis, defective interleukin-1 signalling and lymph node organogenesis in TRAF6-deficient mice. Genes Cells 4:353–362
Neutra MR, Frey A, Kraehenbuhl JP (1996) Epithelial M cells: gateways for mucosal infection and immunization. Cell 86:345–348
Nishikawa S, Honda K, Vieira P, Yoshida H (2003) Organogenesis of peripheral lymphoid organs. Immunol Rev 195:72–80
Okuda M, Togawa A, Wada H, Nishikawa S (2007) Distinct activities of stromal cells involved in the organogenesis of lymph nodes and Peyer's patches. J Immunol 179:804–811
Pappo J, Mahlman RT (1993) Follicle epithelial M cells are a source of interleukin-1 in Peyer's patches. Immunology 78:505–507
Pellas TC, Weiss L (1990) Migration pathways of recirculating murine B cells and CD4+ and CD8+ T lymphocytes. Am J Anat 187:355–373
Sixt M, Kanazawa N, Selg M, Samson T, Roos G, Reinhardt DP, Pabst R, Lutz MB, Sorokin L (2005) The conduit system transports soluble antigens from the afferent lymph to resident dendritic cells in the T cell area of the lymph node. Immunity 22:19–29
Sun Z, Unutmaz D, Zou YR, Sunshine MJ, Pierani A, Brenner-Morton S, Mebius RE, Littman DR (2000) Requirement for RORgamma in thymocyte survival and lymphoid organ development. Science 288:2369–2373
Suzuki K, Ha SA, Tsuji M, Fagarasan S (2007) Intestinal IgA synthesis: a primitive form of adaptive immunity that regulates microbial communities in the gut. Semin Immunol 19:127–135
Tsuji M, Komatsu N, Kawamoto S, Suzuki K, Kanagawa O, Honjo T, Hori S, Fagarasan S (2009) Preferential generation of follicular B helper T cells from Foxp3+ T cells in gut Peyer's patches. Science 323:1488–1492
Tyrer PC, Ruth Foxwell A, Kyd JM, Otczyk DC, Cripps AW (2007) Receptor mediated targeting of M-cells. Vaccine 25:3204–3209
Veiga-Fernandes H, Coles MC, Foster KE, Patel A, Williams A, Natarajan D, Barlow A, Pachnis V, Kioussis D (2007) Tyrosine kinase receptor RET is a key regulator of Peyer's patch organogenesis. Nature 446:547–551
Villena A, Zapata A, Rivera-Pomar JM, Barrutia MG, Fonfria J (1983) Structure of the non-lymphoid cells during the postnatal development of the rat lymph nodes. Fibroblastic reticulum cells and interdigitating cells. Cell Tissue Res 229:219–232
Wijburg OL, Uren TK, Simpfendorfer K, Johansen FE, Brandtzaeg P, Strugnell RA (2006) Innate secretory antibodies protect against natural Salmonella typhimurium infection. J Exp Med 203:21–26
Yamamoto M, Kweon MN, Rennert PD, Hiroi T, Fujihashi K, McGhee JR, Kiyono H (2004) Role of gut-associated lymphoreticular tissues in antigen-specific intestinal IgA immunity. J Immunol 173:762–769
Yokota Y, Mansouri A, Mori S, Sugawara S, Adachi S, Nishikawa S-I, Gruss P (1999) Development of peripheral lymphoid organs and natural killer cells depends on the helix-loop-helix inhibitor Id2. Nature 397:702–706

Yoshida H, Honda K, Shinkura R, Adachi S, Nishikawa S, Maki K, Ikuta K, Nishikawa SI (1999) IL-7 receptor alpha+ CD3(-) cells in the embryonic intestine induces the organizing center of Peyer's patches. Int Immunol 11:643–655

Yoshida H, Naito A, Inoue J, Satoh M, Santee-Cooper SM, Ware CF, Togawa A, Nishikawa S (2002) Different cytokines induce surface lymphotoxin-alphabeta on IL-7 receptor-alpha cells that differentially engender lymph nodes and Peyer's patches. Immunity 17:823–833

Yoshida K, Kaji M, Takahashi T, van den Berg TK, Dijkstra CD (1995) Host origin of follicular dendritic cells induced in the spleen of SCID mice after transfer of allogeneic lymphocytes. Immunology 84:117–126

# Chapter 10
# Cryptopatches and Isolated Lymphoid Follicles: Aspects of Development, Homeostasis and Function

**Heike Herbrand and Oliver Pabst**

**Abstract** Visible by the naked eye Peyer's patches are the most prominent organized lymphoid tissue in the mammalian small intestine. Taking a closer look with a microscope one might be surprised that the intestinal wall in addition to Peyer's Patches harbours a huge number of smaller lymphoid aggregates, which are heterogeneous in size and cellular composition. These range from tiny clusters of lymphoid progenitor cells in the crypt zone, referred to as Cryptopatches, to more robust aggregations termed Isolated lymphoid follicles, that bulge out from the crypt zone and resemble Peyer's patch follicles in aspects of architecture, cellular composition and function. Recent reports revealed that Cryptopatches develop into Isolated lymphoid Follicles in response to adequate extrinsic stimulation. Although most immunologists neglected these lymphoid structures for decades, on a whole they might contain at least as many immune cells as Peyer's Patches and seem to be required for induction of intestinal IgA and maintenance of gut homeostasis. This chapter outlines phenotype, development and putative functions of these small solitary intestinal lymphoid structures.

## 10.1 Cryptopatches and Isolated Lymphoid Follicles Are Differentially Matured Aspects of One Basic Structure

Described already three centuries ago by the Swiss physician Conrad Peyer, Peyer's Patches (PP) are the most prominent organized lymphoid tissue in the small intestine and serve the induction of intestinal immune responses. Besides 8 to 12 PP, the murine intestine contains more than 1,000 small lymphoid aggregates (Kanamori et al.1996, Pabst et al. 2005, Velaga et al. 2009) and even though no careful quantification has been done, numerous small lymphoid aggregates have also been reported in the intestines of human, rabbit, rat and guinea pig (Hitotsumatsu et al. 2005; Keren et al. 1978; Moghaddami et al. 1998; Rosner and Keren 1984).

H. Herbrand (✉) and O. Pabst
Institute of Immunology, Hannover Medical School, Hannover, Germany
e-mail: herbrand.heike@mh-hannover.de

P. Balogh (ed.), *Developmental Biology of Peripheral Lymphoid Organs*,
DOI 10.1007/978-3-642-14429-5_10, 

Based on differences in size and cellular composition as well as on the historical context of their discovery, these structures were assigned different names and putative functions. The smallest structures with an average diameter of roughly 50 μm were termed cryptopatches (CP). CP are confined to the crypt zone of the intestinal epithelium, well hidden by the villi protruding into the gut lumen, and seem to be randomly distributed within the gut wall. They are mainly composed of cells that have not yet been allocated to any hematopoietic cell lineage and express the stem cell factor receptor ckit ($lin^{-}ckit^{+}$ cells). Furthermore, CP contain a substantial proportion of $CD11c^{+}$ dendritic cells (DC) but are almost void of mature B and T lymphocytes (Kanamori et al. 1996). It is the matter of an ongoing debate whether or not CP are breeding places of intraepithelial lymphocytes (IEL), a specialized T cell population that – under adequate experimental conditions – can be generated independent of the thymus (see Sect. 10.6.1).

In 2001, only 5 years after the initial description of CP, Isolated lymphoid follicles (ILF) were reported as another type of lymphoid tissue in the murine small intestine (Hamada et al. 2002). With an average diameter of about 150 μm, ILF are larger than CP. They are dome-shaped structures, broader but shorter than

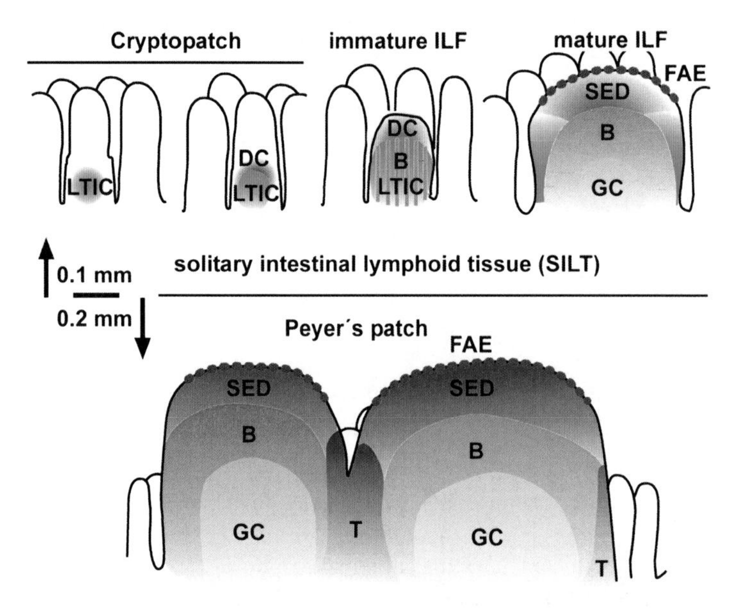


**Fig. 10.1** Schematic comparison of CP/ILF and PP in the murine small intestine. Cryptopatches (CP) and Isolated Lymphoid Follicles (ILF) are differentially matured aspects of one common structure also referred to as solitary intestinal lymphoid tissue (SILT). Via a series of intermediate phenotypes, CP can develop into mature Isolated lymphoid follicles (mature ILF). CP are clusters of LTIC (*green*) and DC (*blue*) located in the crypt zone. Recruitment of mature lymphocytes, mainly B cells (*red*), is resulting in CP to ILF transition. Mature ILF are composed of a compact B cell follicle containing a central germinal centre, a subepithelial dome harbouring $CD11c^{+}$ DC and an M cell sufficient FAE (*thick grey dotted line*). Mature ILF resemble PP follicles but are smaller in size and lack distinct T cell zones. Abbreviations: *LTIC* Lymphoid Tissue Inducer Cells; *DC* Dendritic Cells; *B* B cell follicle; *GC* Germinal Centre; *SED* Subepithelial Dome; *FAE* Follicle-Associated Epithelium; *T* T cell zone

the surrounding villi and bulge out from the crypt zone (Fig. 10.1). In architecture and cellular composition, ILF share important features of PP: They contain a central B cell follicle and are covered by a follicle-associated epithelium harbouring M cells, a cell type specialized for the uptake of antigen from the gut lumen. In contrast to well defined T cell zones present in PP, ILF contain only few T cells that are dispersed throughout the structure and a considerable number of $lin^{-}ckit^{+}$ lymphoid progenitor cells. ILF are capable of generating germinal centres, indicating that IgA class switch and affinity maturation might occur in these structures.

However, a detailed analysis of huge numbers of such small lymphoid aggregations in the small intestine revealed that the majority of structures cannot unambiguously be identified as either CP or ILF but instead display intermediate phenotypes (Pabst et al. 2005). These findings fuelled the hypothesis that CP and ILF are no separate types of structures but instead form the outermost poles of a broad continuum of structures. Indeed, numerous recent reports suggest that ILF develop out of CP in response to adequate extrinsic stimulation.

## 10.2 Gut Bacteria Modulate the CP/ILF Spectrum

The presence of gut commensal bacteria is a prerequisite for the full maturation of gut-associated lymphoid tissue. Mice grown in a sterile environment (germfree mice) lack an intestinal commensal microflora. They are distinguished by hypocellular PP and a CP/ILF spectrum composed of structures that are reduced in size and B cell content and lack germinal centres (Hamada et al. 2002, Pabst et al. 2006). Still, the total number of CP/ILF structures found in germ free mice matches that of conventionally reared mice.

Thus, CP formation is independent of gut bacteria whereas maturation of the structures, i.e. generation of ILF, requires signals deriving from the intestinal microflora. Interestingly, the endogenous time window of ILF formation, 3–4 weeks after birth, coincides with the time of weaning that brings about the uptake of solid food and is accompanied with a boost of bacterial input and antigen challenge.

Colonization of germ-free mice by housing them in the same cage with specific pathogen free (SPF) mice provokes an adaptation of their CP/ILF spectrum to that of SPF mice. Conversely, antibiosis of SPF mice reverses the CP/ILF spectrum to one characteristic of germ free mice (Pabst et al. 2006) indicating that continuous triggers from the endogenous intestinal microflora are required for the development and maintenance of the full CP/ILF spectrum.

How can commensal organisms colonizing the gut lumen influence the maturation of lymphoid structures in the gut epithelium? Recognition of bacteria by the immune system involves the ligation of receptors of microbe-associated molecular patterns of the Toll-like (TLR) and nucleotide-binding oligomerization domain (NOD)-like receptor families. Maturation of ILF is incomplete in mutants lacking members of these receptor families, for example in TLR2/4-, MyD88-, TRIF-, NOD1- or NOD2-deficient mice, whereas the formation of the organ anlagen, i.e. the CP, is unaffected in these mutants (Bouskra et al. 2008). TLR are expressed by

CP/ILF stroma cells and TLR ligation upregulates expression of the adhesion molecule MAdCAM-1 and of chemokines CXCL13, CCL19 and CCL20 that control the homing of immune cells into the growing CP/ILF. Thus, signals derived from commensal bacteria in the gut lumen and from hematopoietic progenitor cells in the CP (see Sect. 10.5) are integrated by CP stroma cells and synergistically induce a set of molecules driving further ILF maturation via the recruitment of DC, B and T lymphocytes. A similar mechanism was deduced from in vitro experiments where the co-culture of lymphoid tissue inducer cells (LTIC, see Sect 10.5) with gut stroma cells provoked the induction of CCL19 and MAdCAM-1. This effect was further enhanced by the addition of TLR-4 ligand lipopolysaccharide (LPS) to the culture medium (Tsuji et al. 2008). CP to ILF transition is blocked in chemokine receptor CCR6-deficient mice (McDonald et al. 2007). As the expression of CCR6 ligand CCL20 is induced by inflammatory signals, this phenotype fits with a role of CCR6/CCL20 in mediating CP to ILF transition in response to microbial/inflammatory signals. Besides CCL20, β-defensin is another ligand of CCR6. Mice lacking the β-defensin mouse homologue (mBD3) show the same defect in ILF formation as CCR6-deficient mice. Blocking of CCR6 ligand CCL20 by antibody treatment similarly impairs the formation of ILF (Bouskra et al. 2008). mBD3 expression on inflamed epithelia and intestinal crypts is induced by NOD1. NOD1 that is expressed by CP/ILF stroma cells recognizes a peptidoglycane substructure primarily contained in the cell wall of gram-negative bacteria (Fritz et al. 2006). Integration of the former data suggests a mechanism of CP to ILF transition as follows: Sensing of gut commensal bacteria by NOD-1 upregulates CCL20 and mBD3 which in turn attract $CCR6^+$ DC and B cells required for the CP to ILF transition. The presence of $CD11c^+$ DC in the CP might be mandatory for ILF development as these DC express a set of receptors required for bacterial sensing.

A recent study analyzed the type of bacteria required for ILF maturation. Germfree mice colonized with gram-positive *Lactobacillae* developed few if any ILF, whereas many ILF formed upon colonization with the gram-negative species *Bacteroides* or *E.coli*. Consistently, newborn mice treated by the antibiotic vanomycin that selectively kills gram-positive bacteria developed significant numbers of ILF whereas only few ILF were found in mice treated with the antibiotic colistin that kills gram-negative bacteria (Bouskra et al. 2008).

Mature ILF contribute to the intestinal IgA pool (see Sect. 10.6.2), which restricts the number of bacteria populating the gut tube. Thus, ILF via the induction of IgA are part of a regulatory feedback loop that controls gut commensal homeostasis and the maturation status of gut-associated lymphoid tissue.

## 10.3 Postnatal Formation of CP/ILF

PP formation occurs during a narrow time window during embryogenesis (see Chap. 9). In contrast, development of CP and ILF occurs postnatally. As mentioned above ILF develop from CP, i.e. CP are the anlagen of ILF. CP can be detected

14–17 days after birth (Kanamori et al. 1996). ILF have been described to appear around day 21–28 (Hamada et al. 2002). The onset of CP to ILF transition thus coincides with the period of weaning characterized by the first uptake of solid food. Confrontation with a variety of food antigens and augmented encounters with intestinal pathogens boost the maturation of the intestinal immune system and are accompanied by an increased density and complexity of commensal microflora.

Formation of PP can be blocked by treatment of pregnant mice with a neutralizing anti-Interleukin-7-receptor antibody or an agonistic Lymphotoxin-βR Ig fusion protein during embryonic days 14.5–17.5. This gestational treatment does, however, not prevent the formation of CP and ILF. In contrast, continuous interference with these signalling pathways in newborn mice results in the absence of CP/ILF. These findings suggest that similar mechanisms acting at distinct time windows might control CP/ILF and PP development.

Although regular development of CP/ILF occurs during the neonatal period, the process – other than PP formation – can be faithfully recapitulated in adult animals supposed that adequate signals are present. The transfer of bone marrow from adult wild-type mice to adult Lymphotoxin α deficient mice that lack CP/ILF was shown to reconstitute the CP/ILF compartment (Lorenz et al. 2003). This suggests that compared with PP organogenesis, CP/ILF formation is more flexible and temporally not restricted to a narrow time window.

## 10.4 The Mechanisms Controlling CP Formation Resemble Those of PP Formation

As described above, ILF develop from CP, i.e. CP represent the organ anlagen of ILF. In this light, strictly speaking, the term organogenesis should be reserved for the formation of CP. In contrast, transition of CP to ILF is achieved by the recruitment of DC and lymphocytes to the CP and should be regarded a process of maturation rather than organogenesis.

Even though secondary lymphoid organs (SLO) considerably differ in location, organ size and cellular composition, a similar set of molecules and signalling cascades are required for their generation (Vondenhoff et al. 2007). SLO organogenesis is initiated by the immigration of a unique population of LTIC deriving from the fetal liver to the future organ site. Interaction of these LTIC with local stroma cells leads to the clustering of LTIC thereby forming the SLO anlage. Numerous molecules have been identified mediate the interaction of LTIC with stromal organizer cells. Finally, a positive feedback loop is established that promotes the growth of the organ anlage by attracting further LTIC and, subsequently, DC and mature lymphocytes (see Chap. 7).

Although CP/ILF are generated postnatally, evidence accumulates that their formation requires a similar set of molecules as the organogenesis of classical SLO during gestation. A prerequisite of CP formation is the presence of LTIC-like ckit$^+$lin$^-$ cells. Consequently, the lack of molecules required for the development of LTIC-like cells will result in the absence of CP. Development of ckit$^+$lin$^-$ CP cells has been shown to depend on the γt isoform of the retinoic acid related orphan receptor (RORγt) and on the transcription factor Inhibitor of differentiation 2 (Id2). Mice deficient for RORγt or Id2 lack any SLO except of the spleen (Sun et al. 2000, Yokota et al. 1999) and, moreover, they lack ckit$^+$lin$^-$ cells in the intestinal mucosa as well as CP. CP form by clustering of LTIC-like ckit$^+$lin$^-$ cells within the intestinal epithelium. Similar to the formation of the LN and PP anlagen, this process requires the interaction of LTIC like cells with local stromal organizer cells. These are mediated by binding of LTα1β2 expressed by the hematopoietic cells to LTβR on the surface of the stroma cells. Induction of LTα1β2 on LTIC cells depends on Interleukin 7 (IL-7) signalling. Consequently, mice deficient for IL-7Rα, LTα, LTβ or the LT-βR do harbour ckit$^+$lin$^-$ cells in the small intestinal epithelium. Yet these progenitor cells fail to cluster to form CP but instead remain randomly distributed within the lamina propria. Besides the membrane-bound LTα1β2, LT also exists in a secreted form that is composed of three LTα subunits. In contrast to LTα1β2, this LTα3 homotrimer does not ligate LTβR but signals via Tumor necrosis factor-receptor-I (TNFR-I) that is expressed by stroma cells of the developing CP/ILF. TNFR-I deficient mice lack mature ILF while PP formation is not affected. It is yet unclear, at which stage CP/ILF development is blocked in these mutants (Lorenz et al. 2003).

Molecules downstream in the LT signalling cascade are required for the CP to ILF transition and full maturation of ILF. LTβR downstream signalling involves two pathways: the canonical pathway requires IKKα, whereas the non-canonical pathway involves NF-κb inducing Kinase (NIK). Both pathways finally merge in the activation of nuclear factor κb (NFκb) that induces the production of homeostatic chemokines such as CCL21, CCL19 and CXCL13. NIK is expressed on VCAM-1$^+$ LTβR$^+$ stroma cells in the developing CP/ILF and CP of Map3K14$^{aly/aly}$ mice that lack a functional NIK are less compact and almost void of CD11c$^+$ DC and mature B cells. This phenotype might be caused by the downregulation of adhesion molecules VCAM-1 and MAdCAM-1 and chemokines CXCL13 and CCL19 (Tsuji et al. 2008). In summary, LTβR signals mediating the clustering of LTIC at the early phase of CP formation require the canonical NFκb-pathway. Later on the non-canonical NFκb pathway involving NIK is involved in controlling the CP to ILF transition.

In general, the CP to ILF transition is mainly achieved by the recruitment of different cell types – mainly DC, B and T cells – to the CP. Consequently, molecules controlling the migration of immune cells, like integrins and adhesion molecules as well as chemokines and their receptors, are the key players in this process. The integrins α4β1 and α4β7 are expressed by LTIC like CP cells and, moreover by mature lymphocytes in the ILF, while CP stroma cells express VCAM-1. β7 integrin deficient mice develop DC sufficient CP, whereas the immigration of B and T cells is blocked (Wang et al. 2008). At least two chemokine receptors, CCR6 and CXCR5, control the homing of B cells to the developing ILF. Both of them are expressed by the majority of LTIC like cells in CP and ILF as well as by mature ILF B cells (McDonald et al. 2007; Velaga 2009). In mice deficient for CCR6 CP formation is intact, whereas the number of B cell sufficient ILF is severely reduced. Similarly, CCR6 deficient mice harbour a lower number of B cell follicles in their PP. In mice lacking CXCR5, the onset of CP formation is delayed. This defect is partially compensated until the age of 8 weeks by the generation of aberrantly enlarged but B cell deficient CP/ILF-like structures. The molecular mechanism that facilitates CP formation independent of CXCR5 remains unknown. As CCR7/CXCR5 double deficient mice possess normal CP and ILF numbers, this compensatory mechanism apparently does not involve chemokine receptor CCR7, which is also expressed by LTIC like cells (Velaga et al. 2009).

Moreover, recent data hint to synergistic rather than redundant functions of CXCR5 and CCR6 at early stages of ILF development. CXCL13 but not CCL20 was shown to induce the expression of Lymphotoxin in the early phase of LTIC clustering, while CCL20 but not CXCL13 expression in the intestinal epithelium was found to be upregulated in response to inflammatory stimuli, which contribute to the full maturation of ILF (see Sect. 10.6).

These defects in ILF and PP development might explain why CCR6 deficient mice generate impaired mucosal immune responses (Cook et al. 2000; Velaga et al. 2009).

A role of chemokine receptor CCR9 in CP formation was implicated from the analysis of mice that express a CCL25 intrakine gene preventing CCR9 expression on the cell surface and possess few and small CP (Onai et al. 2002). Still, the observation that CCR9 deficiency does not affect CP/ILF formation (O. Pabst, unpublished observation) argues against a prominent role of CCR9 in CP/ILF development.

## 10.5 Putative Functions of CP/ILF

CP/ILF have only recently been investigated in more detail and the function of these structures is still poorly understood. Generally, CP/ILF have been implicated into two processes, the local thymus-independent generation of intraepithelial T lymphocytes and the generation of intestinal IgA responses. Evidence supporting or contradicting either function is briefly discussed in this section.

### *10.5.1 lin$^-$ckit$^+$ CP Cells: T Cell Progenitors or Organizer Cells Required for ILF Maturation?*

Initially, lin$^-$ckit$^+$ CP cells were regarded as progenitors of gut IEL (Kanamori et al. 1996). Under appropriate experimental conditions IEL are capable of developing independent of the thymus and were assumed to differentiate locally in the gut. Supportive of this, the transfer of $lin^-ckit^+$ CP cells into irradiated severe combined immuno deficient (SCID) mice that lack any B and T lymphocytes was shown to reconstitute the IEL population in the mutants' intestines (Saito et al. 1998). On the other hand, Lymphotoxin deficient mice lack CP but still possess normal numbers of IEL (Pabst et al. 2005) indicating that generation of IEL might not be the main function of $lin^-ckit^+$ CP cells. Other studies suggested that $lin^-ckit^+ROR\gamma t^+$ CP cells represent adult counterparts of LTIC and act as local organizers of mucosal lymphoid tissue (Eberl 2005). As described in Sect. 10.5, co-cultivation of gut stroma cells with adult $ROR\gamma t^+$ LTIC upregulates molecules involved in ILF maturation (Tsuji et al. 2008). This indicates that $lin^-ckit^+$ CP cells indeed regulate CP/ILF homeostasis.

Eventually, $lin^-$ CP cells might be involved in both processes. A detailed phenotypical analysis of $lin^-$ CP cells revealed that these cells are highly heterogeneous. Unexpectedly, only a minority of cells displayed an LTIC like $ROR\gamma t^{high}$ phenotype while CP cells with intermediate or low $ROR\gamma t$ levels expressed markers indicative of the commitment to the T cell lineage (Naito et al. 2008). Thus, an LTIC-like subpopulation of CP cells might control ILF formation and maturation whereas a $ROR\gamma t$-negative subpopulation might represent T cell precursors capable of contributing to the intestinal IEL pool.

### 10.5.2 The Putative Role of ILF in Intestinal IgA Immunity

PP have since long been recognized as prominent inductive sites for antigen-specific IgA responses. Mature ILF closely resemble PP follicles in architecture and cellular composition in that (1) they possess a central B cell follicle capable of forming germinal centres, (2) they are covered with a follicle-associated epithelium containing M cells specialized for the uptake of material from the gut lumen. (3) They contain a network of FDCs which is a prerequisite of GC formation. (4) They express Activation Induced cytidine Deaminase (AID) a molecule crucially required for Ig class switching. This similarity, together with the rapid adaptation of the ILF phenotype to challenges arising from gut microbiota, strongly suggests a role of ILF in intestinal IgA responses. Mice rendered PP-deficient by gestational LTβR-Ig treatment were shown to mount an antigen specific IgA response when orally challenged with the soluble proteins Ovalbumin and cholera toxin. Similarly, LTα-deficient mice reconstituted with wild-type bone marrow to contain CP/ILF but no PP generated an antigen specific IgA response after oral infection with enteropathogenic Salmonellae, suggesting the existence of a PP-independent pathway for antigen specific intestinal IgA responses (Yamamoto et al. 2000; Lorenz 2004). CP/ILF were shown to represent major sites of *Salmonella* invasion and ensuing mucosal pathology also in the presence of PP (Halle et al. 2007). Furthermore, ILF and PP harbour a similar proportion of $IgM^+B220^+$ and $IgA^+B220^+$ B cells, whereas the lamina propria only contains plasma blasts and plasma cells that have already undergone IgM to IgA class switch, suggesting that ILF could indeed provide inductive sites for the generation of IgA-committed B cells (Shikina et al. 2004). Two comprehensive studies that were recently published however put doubt on a crucial role of intestinal lymphoid aggregations other than PP in the generation of intestinal IgA responses towards T cell-dependent antigens. ILF sufficient mice that lack PP and MLN failed to mount an antigen specific IgA response upon oral immunization with a recombinant Salmonella strain expressing the Tetanus toxin C fragment (Hashizume et al. 2008). Similarly, CXCR5 deficient mice reconstituted with wild-type bone marrow possess a wild-type like CP/ILF spectrum but still fail to mount an effective IgA response to orally applied cholera toxin or following oral infection with Salmonella (Velaga et al. 2009). Although presumably non-involved in T-dependent IgA responses, there is evidence that ILF are important for the generation of T cell independent intestinal immune responses. Mice that lack any T cells cannot form germinal centres but still possess considerable numbers of $IgA^+$ plasma cells in the lamina propria that develop in the ILF but not in the PP (Tsuji et al. 2008). A role of ILF in T-independent IgA responses fits well with the more simple architecture of ILF that lack a concrete T cell zone rich in DC, which is a characteristic of PP and a prerequisite of T cell-dependent immune responses.

As discussed in Sect. 10.3, intestinal microbiota are crucial for the formation of mature ILF. In turn, the presence of mature ILF – via IgA production – largely impacts the number and composition of the gut commensal population. This interrelation was first noticed, when it turned out that mice deficient for AID that lack IgA due to a defect in class switch recombination develop hyperblastic but B cell deficent ILF. Simultaneously, AID-deficiency leads to an expansion of anaerobic bacteria in the small intestine. Antibiosis of AID-deficient mice reduces the bacterial load of the gut and coevally abolishes CP/ILF hyperplasia. Similar does the normalization of IgA titers by parabiosis of an AID-defcient mouse with an IgA sufficient wild-type mouse (Fagarasan et al. 2002; Suzuki et al. 2004). Moreover, gut bacteria were found to be strongly expanded in mice deficient for NOD-1 or mBD3 that lack mature ILF. The same phenotype results from the ablation of CP/ILF by postnatal treatment with LTβR-Ig fusion protein (Bouskra et al. 2008). Thus ILF are crucially involved in controlling the hosts load with intestinal microbiota, most likely via the production of IgA.

## 10.6 Why CP/ILF Should Be Regarded as Classical Secondary Lymphoid Organs

Lymph nodes and PP are classical SLO. Their position and number are determined during gestation, and their organogenesis does not require activating environmentally stimuli. Therefore, postnatal events may alter parameters of the phenotypic appearance like size or cellular composition, but will not affect their location or number. In contrast, tertiary lymphoid organs (TLO) can form de novo at any time an appropriate stimulation occurs in a process termed lymphoid neogenesis. In contrast to SLO, not all TLO seem to require the presence of LTIC for their formation (Marinkovic et al. 2006; Moyron-Quiroz et al. 2004).

How should CP/ILF be classified within this field? Although maturation of CP depends on appropriate stimulation and therefore phenotypical appearance of CP/ILF is exceedingly plastic, their total number is constant. Moreover, CP/ILF – once established during postnatal development – do not regularly form de novo and disappear (Velaga et al. 2009). Therefore, formation of CP/ILF shares decisive features of SLO rather than TLO, and these structures should be regarded as classical lymphoid organs and integral part of the gut-associated lymphoid tissue.

## References

Bouskra D, Brezillon C, Berard M, Werts C, Varona R, Boneca IG, Eberl G (2008) Lymphoid tissue genesis induced by commensals through NOD1 regulates intestinal homeostasis. Nature 456:507–510

Cook DN, Prosser DM, Forster R, Zhang J, Kuklin NA, Abbondanzo SJ, Niu XD, Chen SC, Manfra DJ, Wiekowski MT, Sullivan LM, Smith SR, Greenberg HB, Narula SK, Lipp M, Lira SA (2000) CCR6 mediates dendritic cell localization, lymphocyte homeostasis, and immune responses in mucosal tissue. Immunity 12:495–503

Eberl G (2005) Inducible lymphoid tissues in the adult gut: recapitulation of a fetal developmental pathway? Nat Rev Immunol 5:413–420

Fagarasan S, Muramatsu M, Suzuki K, Nagaoka H, Hiai H, Honjo T (2002) Critical roles of activation-induced cytidine deaminase in the homeostasis of gut flora. Science 298:1424–1427

Fritz JH, Ferrero RL, Philpott DJ, Girardin SE (2006) Nod-like proteins in immunity, inflammation and disease. Nat Immunol 7:1250–1257

Halle S, Bumann D, Herbrand H, Willer Y, Dahne S, Forster R, Pabst O (2007) Solitary intestinal lymphoid tissue provides a productive port of entry for Salmonella enterica serovar Typhimurium. Infect Immun 75:1577–1585

Hamada H, Hiroi T, Nishiyama Y, Takahashi H, Masunaga Y, Hachimura S, Kaminogawa S, Takahashi-Iwanaga H, Iwanaga T, Kiyono H, Yamamoto H, Ishikawa H (2002) Identification of multiple isolated lymphoid follicles on the antimesenteric wall of the mouse small intestine. J Immunol 168:57–64

Hashizume T, Togawa A, Nochi T, Igarashi O, Kweon MN, Kiyono H, Yamamoto M (2008) Peyer's patches are required for intestinal immunoglobulin A responses to Salmonella spp. Infect Immun 76:927–934

Hitotsumatsu O, Hamada H, Naganuma M, Inoue N, Ishii H, Hibi T, Ishikawa H (2005) Identification and characterization of novel gut-associated lymphoid tissues in rat small intestine. J Gastroenterol 40:956–963

Kanamori Y, Ishimaru K, Nanno M, Maki K, Ikuta K, Nariuchi H, Ishikawa H (1996) Identification of novel lymphoid tissues in murine intestinal mucosa where clusters of c-kit+ IL-7R+ Thy1+ lympho-hemopoietic progenitors develop. J Exp Med 184:1449–1459

Keren DF, Holt PS, Collins HH, Gemski P, Formal SB (1978) The role of Peyer's patches in the local immune response of rabbit ileum to live bacteria. J Immunol 120:1892–1896

Lorenz RG, Chaplin DD, McDonald KG, McDonough JS, Newberry RD (2003) Isolated lymphoid follicle formation is inducible and dependent upon lymphotoxin-sufficient B lymphocytes, lymphotoxin beta receptor, and TNF receptor I function. J Immunol 170:5475–5482

Marinkovic T, Garin A, Yokota Y, Fu YX, Ruddle NH, Furtado GC, Lira SA (2006) Interaction of mature CD3+CD4+ T cells with dendritic cells triggers the development of tertiary lymphoid structures in the thyroid. J Clin Invest 116:2622–2632

McDonald KG, McDonough JS, Wang C, Kucharzik T, Williams IR, Newberry RD (2007) CC chemokine receptor 6 expression by B lymphocytes is essential for the development of isolated lymphoid follicles. Am J Pathol 170:1229–1240

Moghaddami M, Cummins A, Mayrhofer G (1998) Lymphocyte-filled villi: comparison with other lymphoid aggregations in the mucosa of the human small intestine. Gastroenterology 115:1414–1425

Moyron-Quiroz JE, Rangel-Moreno J, Kusser K, Hartson L, Sprague F, Goodrich S, Woodland DL, Lund FE, Randall TD (2004) Role of inducible bronchus associated lymphoid tissue (iBALT) in respiratory immunity. Nat Med 10:927–934

Naito T, Shiohara T, Hibi T, Suematsu M, Ishikawa H (2008) ROR gamma t is dispensable for the development of intestinal mucosal T cells. Mucosal Immunol 1:198–207

Onai N, Kitabatake M, Zhang YY, Ishikawa H, Ishikawa S, Matsushima K (2002) Pivotal role of CCL25 (TECK)-CCR9 in the formation of gut cryptopatches and consequent appearance of intestinal intraepithelial T lymphocytes. Int Immunol 14:687–694

Pabst O, Herbrand H, Worbs T, Friedrichsen M, Yan S, Hoffmann MW, Korner H, Bernhardt G, Pabst R, Forster R (2005) Cryptopatches and isolated lymphoid follicles: dynamic lymphoid tissues dispensable for the generation of intraepithelial lymphocytes. Eur J Immunol 35:98–107

Pabst O, Herbrand H, Friedrichsen M, Velaga S, Dorsch M, Berhardt G, Worbs T, Macpherson AJ, Forster R (2006) Adaptation of solitary intestinal lymphoid tissue in response to microbiota and chemokine receptor CCR7 signaling. J Immunol 177:6824–6832

Rosner AJ, Keren DF (1984) Demonstration of M cells in the specialized follicle-associated epithelium overlying isolated lymphoid follicles in the gut. J Leukoc Biol 35:397–404

Saito H, Kanamori Y, Takemori T, Nariuchi H, Kubota E, Takahashi-Iwanaga H, Iwanaga T, Ishikawa H (1998) Generation of intestinal T cells from progenitors residing in gut cryptopatches. Science 280:275–278

Shikina T, Hiroi T, Iwatani K, Jang MH, Fukuyama S, Tamura M, Kubo T, Ishikawa H, Kiyono H (2004) IgA class switch occurs in the organized nasopharynx- and gut-associated lymphoid tissue, but not in the diffuse lamina propria of airways and gut. J Immunol 172:6259–6264

Sun Z, Unutmaz D, Zou YR, Sunshine MJ, Pierani A, Brenner-Morton S, Mebius RE, Littman DR (2000) Requirement for RORgamma in thymocyte survival and lymphoid organ development. Science 288:2369–2373

Suzuki K, Meek B, Doi Y, Muramatsu M, Chiba T, Honjo T, Fagarasan S (2004) Aberrant expansion of segmented filamentous bacteria in IgA-deficient gut. Proc Natl Acad Sci USA 101:1981–1986

Tsuji M, Suzuki K, Kitamura H, Maruya M, Kinoshita K, Ivanov II, Itoh K, Littman DR, Fagarasan S (2008) Requirement for lymphoid tissue-inducer cells in isolated follicle formation and T cell-independent immunoglobulin A generation in the gut. Immunity 29:261–271

Velaga S, Herbrand H, Friedrichsen M, Jiong T, Dorsch M, Hoffmann MW, Forster R, Pabst O (2009) Chemokine receptor CXCR5 supports solitary intestinal lymphoid tissue formation, B cell homing, and induction of intestinal IgA responses. J Immunol 182:2610–2619

Vondenhoff MF, Kraal G, Mebius RE (2007) Lymphoid organogenesis in brief. Eur J Immunol 37 (Suppl 1):S46–S52

Wang C, McDonough JS, McDonald KG, Huang C, Newberry RD (2008) Alpha4beta7/MAdCAM-1 interactions play an essential role in transitioning cryptopatches into isolated lymphoid follicles and a nonessential role in cryptopatch formation. J Immunol 181:4052–4061

Yamamoto M, Rennert P, McGhee JR, Kweon MN, Yamamoto S, Dohi T, Otake S, Bluethmann H, Fujihashi K, Kiyono H (2000) Alternate mucosal immune system: organized Peyer's patches are not required for IgA responses in the gastrointestinal tract. J Immunol 164:5184–5191

Yokota Y, Mansouri A, Mori S, Sugawara S, Adachi S, Nishikawa S, Gruss P (1999) Development of peripheral lymphoid organs and natural killer cells depends on the helix-loop-helix inhibitor Id2. Nature 397:702–706

# Part IV
# Single Complexity: The Spleen

The spleen in both man and rodents is the largest filter of blood in the body. This filtering function is coupled to the three main activities of the spleen: removal of aging red blood cells, participation in innate immunity, and promoting various forms of adaptive immune responses against blood-borne pathogens. These functions are attributed to three main tissue regions within the organ. Of these, the red pulp filters blood and removes senescent erythrocytes and other effete blood cells; the white pulp and marginal zone represent the lymphoid region of the spleen, and consist of B- and T-lymphocyte-rich lymphoid compartments. Their function to establish adaptive immune responses is based on a highly efficient and dynamic co-operation between its cellular constituents, including both migratory hemopoietic cells and sessile stromal elements.

Of the peripheral lymphoid organs, first the spleen developed during evolution in vertebrates. The presence of spleen in vertebrate classes lacking lymph nodes (from fish through birds) also indicates that, for the overwhelming majority of animals in the biosphere with adaptive immunity, the spleen (along with mucosal lymphoid structures of various degrees of complexity) is sufficient for providing appropriate systemic protection against pathogens. Surprisingly, its main organization scheme has remained largely the same throughout the phylogeny – it is a single organ containing two strikingly different domains, which form during divergent developmental pathways. Generally, its erythro-myelopoietic red pulp part evolves earlier during embryogenesis, and subsequent specification directs the formation of the white pulp, comprised overwhelmingly of lymphoid cells and, to a lesser extent, dendritic cells and macrophages. During evolution the variations in size, shape and position of spleen become more restricted as the animal approaches the class of mammalians where, nevertheless, significant differences exist between various species. This chapter summarizes the main structural and ontogenic characteristics of spleen in various vertebrate classes, and provides a detailed analysis of the developmental properties of spleen in mouse and man. These include both the formation of spleen as a whole organ and its different compartments. Importantly, although several common morphogenic regulators involved in the formation of lymph nodes and Peyer's patches (such as various members of the lymphotoxin/ tumor necrosis family and their downstream signaling components) also participate in the organogenesis of spleen, crucial differences exist between the development of spleen and other peripheral lymphoid organs, including the transcriptional regulation of spleen as a solitary organ and lymph nodes. Finally, the role of organ development and proper splenic tissue compartmentalization in the maintenance of both local and systemic immunological responsiveness will be described in this part.

# Chapter 11
# Structural Evolution of the Spleen in Man and Mouse

**Péter Balogh and Árpád Lábadi**

**Abstract** Of all the peripheral lymphoid tissues, spleen represents the earliest organ specialized to perform adaptive immune responses, and it has also preserved its basic structural organization in various vertebrate classes. Its architecture, as well as developmental properties, is strikingly different from other secondary lymphoid organs. Here we summarize the developmental regulation of its main tissue compartments in man and mouse during the prenatal and early postnatal period, with particular emphasis on its nonhemopoietic stromal constituents and their macrodomains.

## 11.1 Ontogeny of the Spleen in Nonmammalian Vertebrates

In cyclostomes and lungfish, a diffuse lymphoid accumulation in the intestines or stomach wall indicates the earliest type of spleen formation, referred to as prespleen (Tischendorf 1985). In cartilaginous fish (sharks, rays) and teleost fishes, the spleen is an isolated organ situated within the abdominal cavity, although the splenic capsule may also enclose some pancreatic tissue (Fänge and Nilsson 1985). Initiating as a mesenchymal clustering of cells in the left anterior gut, the spleen in fish embryos is largely an erythropoietic organ. The beginning of organ formation is indicated by the expression of Hox11/Tlx1 homeodomain transcription factor, a cardinal regulator in splenic anlage (see below). The organized pattern of spleen is established after hatching, as evaluated by enzymocytochemical staining pattern of its hemopoietic cells (Langenau et al. 2002). Reminiscent of mammalian spleen, the lymphocytes arranged in ellipsoid-like structures around arterioles form the white pulp, where the ellipsoid is surrounded by histiocytes and macrophages. Other myeloid cells, such as granulocytes, are distributed more peripherally from the ellipsoids. Although largely separated, the white pulp in teleost fishes may include

P. Balogh (✉) and A. Lábadi
Department of Immunology and Biotechnology, Faculty of Medicine, University of Pécs, Pécs, Hungary
e-mail: peter.balogh@aok.pte.hu

P. Balogh (ed.), *Developmental Biology of Peripheral Lymphoid Organs*,
DOI 10.1007/978-3-642-14429-5_11, © Springer-Verlag Berlin Heidelberg 2011

erythro-myelopoietic regions. Lymphoid cells accumulate shortly after hatching; however, their lineage affiliation remains controversial, largely owing to the shortage of suitable reagents for distinguishing these cells (Zapata et al. 2006).

Mature B lymphocytes with surface Ig expression appear somewhat later (Hansen and Zapata 1998) and may engage in protective immune responses, including the formation of melano-macrophage centers (MMCs). These structures are presumed piscine analogues of mammalian germinal centers, where the dark color of MMCs derives mostly from melanin, lipofuschsin, and hemosiderin (Vigliano et al. 2006). MMCs contain macrophages intermingled with lymphocytes and may serve as sequestration sites for intracellular bacteria during chronic inflammations. Their hemosiderin content may reflect the erythrophagocytic removal of effete red blood cells and the storage of released iron upon red cell destruction by red pulp macrophages (Agius and Roberts 2003).

In amphibians (Xenopus), the spleen development initiates in the larval period. Originating from the thickening of mesogastrium at around day 4–5 postfertilization (corresponding to 2–3 days after hatching), the splenic anlage initially contains hemopoietic cells only, with the B cell formation residing in the liver. Once separated from the gastrointestinal tract at around day 12, the spleen may engage in immune responses and continues to enlarge throughout the tadpole stage of development. During metamorphosis, a substantial atrophy of spleen occurs, which is followed by the expansion of lymphoid precursors and an increased RAG expression, indicating that the spleen in adult frogs functions both as a primary B cell differentiation site and as a peripheral lymphoid organ. In adult frog spleen, the red pulp and white pulp are clearly separate, with distinct T and B cell rich areas situated in the latter compartment. B cells expressing IgM typically accumulate around arterioles, where they are surrounded by T cells identifiable by CD8 labeling (Du Pasquier et al. 2000; Robert and Ohta 2009).

Although immune responses against bacterial and viral pathogens result in noticeable structural alterations in frog spleen including cell proliferation and formation of plasma cells, no germinal centers are generated. The upregulation of AID during immune responses as a means to alter the effector functions of immunoglobulins is first observed in frogs (Zhao et al. 2005; Marr et al. 2007) and results in isotype switch from IgM to IgX or IgY (analogues of the mammalian IgA and IgG/E, respectively), but the degree of affinity maturation in frogs is substantially inferior compared with species with typical germinal centers, such as birds or mammalians (Flajnik 2002). Other immunoglobulins with lesser known functions produced by frog B cells also include IgD and IgF (Zhao et al. 2006). Due to their infrequent use as subjects in immunological studies, only sparse information is available concerning the developmental and structural properties of spleen in various reptile species.

In birds, the lymph nodes are still absent; although postnatally birds can develop tonsils along the gastrointestinal tract (Del Cacho et al. 1993; Nagy et al. 2005; Nagy and Oláh 2007). Therefore, the spleen still has a dominant role in generating immune responses. The distinctive structural feature of their spleen is the prominence of ellipsoids, which form a compartment separate from the T cell rich periarteriolar lymphoid sheath (PALS). Ellipsoids can also be found in fish and some mammalian spleens. In contrast to fish where macrophage-like cells associating with these structures accumulate and transport antigens to the MMCs in the red pulp, in birds the antigen is transported by ellipsoid-associated cells (EACs) to the periellipsoidal white pulp, a functional analogue to the mammalian marginal zone (MZ) rich in B cells (Jeurissen 1993; Del Cacho et al. 1995).

The core of ellipsoids is formed by penicilliform capillaries, the terminal segments of arteriolar branches originating from the central arterioles. The endothelial cells that cover these capillaries are supported by a thick, yet discontinuous, basement membrane which, in turn, is surrounded by several layers of reticular supporting cells, intermingled with EACs. These EACs may detach and migrate into the PALS or into germinal centers, although their subsequent developmental commitment into diverse antigen-presenting cell subsets is still controversial (Igyártó et al. 2007).

A major difference in the onset of humoral immune responses in birds compared with lower vertebrates is the avian spleen's capacity to establish germinal centers, permitting efficient affinity maturation and isotype switch of immunoglobulins. Subsequent to antigenic challenge, splenic germinal centers will form and organize into dark zone (DZ) and light zone (LZ). In mammalian (primarily human and rodent) germinal centers, DZ and LZ as two adjacent follicular compartments are encapsulated by the lymphocytic corona formed by the remnants of the primary follicle. In birds, however, the LZ occupies a central core, whereas the DZ with rapidly proliferating centroblasts forms an intermediate circumferential ring within a multilayered structure that encloses the central arteriole by a T cell rich external layer (Yasuda et al. 1998).

Developmental studies in chicken spleens reveal that ellipsoids do not appear before embryonic day (ED) 18 of the 21 days prehatching period (Mast and Goddeeris 1999). At day 1, posthatching T cells accumulate around arterioles and also in the red pulp, and by day 7, a substantial T cell enrichment is observed around the arterioles. B cell colonization is biphasic, with a transient drop of B cell number around the hatching. Subsequently, B cells accumulate in the periellipsoidal white pulp, where gradually they are almost completely encircled by macrophages by 7–12-day posthatching. Interestingly, avian spleen also produces homeostatic chemokines, such as CCL19, CCL21, and CXCL13, which are largely under the control of TNF/LT ligands and their receptors in mammalians (Wang et al. 2005). Chickens, however, do not express TNF or lymphotoxin, although an analogue for TNFSF15 (chicken TL1A) has been identified (Takimoto et al. 2005). The identity and role (if any) of various analogues of other members of TNF/LT superfamily in the avian spleen organogenesis remain to be determined. In addition, the gene encoding IL-7 as an important peripheral T cell survival factor (produced by T-zone fibroblastic reticular cells in mammalians) is also present in the chicken genome (Kaiser et al. 2005). Perhaps, the ability to secrete homeostatic chemokines contributes to the chicken spleen's improved capacity to produce high-affinity immunoglobulins during germinal center reaction, although the cellular sources for these chemokines (follicular dendritic cells for CXCL13 and fibroblastic reticular cells for CCL19 and CCL21, respectively both in humans and mice) or IL-7 in chickens have not been formally established yet.

## 11.2 Basic Structural Components of Spleen in Man and Mouse

Despite several common properties of function and structure, there are important anatomical and developmental differences between the spleen of humans and rodents (Steiniger and Barth 2000). Thus, caution is required when extrapolating results from one species to another. In addition, substantial differences between human samples were noted, in contrast to the relative uniform appearance of rodent spleen structure under normal conditions. The aim of this section is to discuss the tissue properties and, where necessary, to point out the differences between the human and rodent spleen architecture. The structural characteristics of chicken, mouse, and human spleen are depicted in Fig. 11.1.

The main scaffolding structure around which the lymphoid architecture of spleen in both species is arranged is formed jointly by the splenic *vasculature* and the surrounding *fibroblastic meshwork*. Together with the extracellular matrix, the

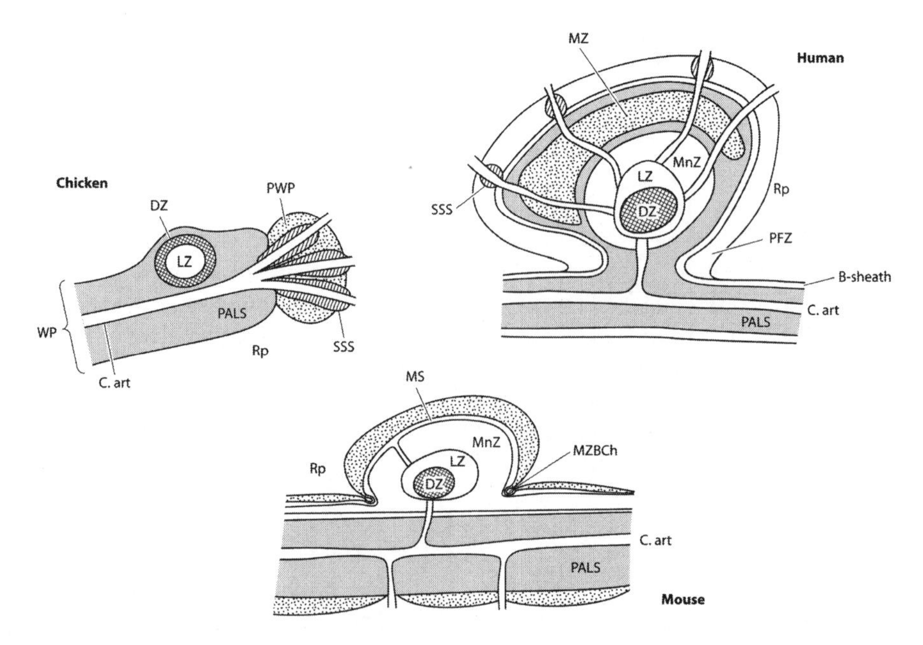


**Fig. 11.1** Main lymphoid compartments of spleen white pulp in chicken [Ch], mouse [Mo], and man [Hu]. Structures with comparable cellular composition and structural properties are indicated in the same shade. B-sheath, a near-continuous layer of B cells positioned extrafollicularly [Hu]; *C. art* central arteriole; *DZ* dark zone of germinal center; *LZ* light zone of germinal center; *MnZ* mantle zone; *MZ* marginal zone; *MZBCh* marginal zone bridging channel [Mo]; *PALS* periarteriolar lymphoid sheath; *PFZ* perifollicular zone [Hu]; *PWP* periellipsoidal white pulp [Ch]; *SSS* Schweigger-Seidel sheath [Ch & Hu]; *Rp* red pulp; *WP* white pulp

reticular meshwork creates the *conduit system*, an intricate network for cell-free fluid transport (see below).

After a certain number of divisions from the splenic artery, the resulting small-sized arterioles (termed central arterioles) are surrounded by the PALS comprised mainly by T cells, whereas the B cell rich follicles are relatively hypovascular structures, although smaller arteriolar branches or capillaries may incidentally traverse them. These two main compartments form the white pulp, which is separated from – or, connected to – the red pulp by the MZ. This region is substantially different between human and rodent species (Mebius et al. 2004). In humans, the MZ contains marginal sinus cells lined by MAdCAM-1 positive cells (Kraal et al. 1995), whereas this structure is absent in mice. The MZ in mouse contains a distinctive set of B cells (MZ B cells) and macrophage subpopulations (Martin and Kearney 2002). MZ macrophages include at least two main subsets as evaluated by their expression of sialoadhesin/Sn/CD169, SIGN-R1, and MARCO antigens as pattern recognition receptors and their differential dependence on macrophage colony-stimulating factor (M-CSF). In mice, the Sn-positive macrophages are located in the proximity of white pulp, whereas MZ macrophages expressing SIGN-R1 receptor are positioned more peripherally, with variable MARCO receptor expression. A smaller subset of MARCO$^{+}$ cells are present in osteopetrosis (*op*/*op*) mice with a mutation in the coding region of M-CSF, whereas SIGN-R1$^{+}$ and Sn$^{+}$ subpopulations are both absent (Ito et al. 1999). Thus in mice there are M-CSF-dependent (Sn$^{+}$/SIGN-R1$^{-}$/MARCO$^{+/-}$ and Sn$^{-}$/SIGN-R1$^{+}$/MARCO$^{+}$) macrophages and also Sn$^{-}$/SIGN-R1$^{-}$/MARCO$^{+}$ macrophages independent from M-CSF. In humans, the CD169$^{+}$ macrophages are localized to the perifollicular zone, a separate region absent in both mouse and rat (Steiniger et al. 1997). Based on the presence of MZ B cells as a distinct B cell subset, human peripheral lymph nodes and Peyer's patches also contain a MZ-analogue region (Tierens et al. 1999), whereas in rodents such compartments have not been identified outside the spleen.

The smallest branches of the central arterioles cross the white pulp and may either terminate in the marginal sinus surrounding the white pulp (in rodents) or may channel to the red pulp sinuses in the perifollicular zone at the red pulp aspect of the MZ (in humans). In humans – with no marginal sinus – some arterioles terminate in a perifollicular zone, while others reach the red pulp, where they connect with collecting venules. The terminal segments of some arteriolar capillaries within the red pulp are surrounded by a complex multilayer structure of ellipsoid (or Schweiger-Seidel sheaths) composed of macrophages, dendritic cells, and a fibroblastic reticular meshwork (Steiniger and Barth 2000), and the endothelial cells in this region also display a considerable heterogeneity, similarly to mouse spleen (Steiniger et al. 2007; Balázs et al. 2001).

In addition to the blood vasculature, the splenic white pulp in some mammalian species contains efferent lymphatic vessels, as evidenced by the electron microscopic description of thin-walled vessels lacking basement membrane. However, little is known about the physiological importance of these vessels in the fluid homeostasis or leukocyte recirculation (Pellas and Weiss 1990).

The *mesenchymal scaffolding* that surrounds the vasculature and forms a three-dimensional lattice throughout the organ is composed of smooth muscle cells, myofibroblasts and reticular fibroblasts, and the extracellular matrix components these cells produce. Previously considered rather passive structural elements of tissue architecture, the fibroblastic reticular cells (FRCs) play important role in sustaining the ordered migration of lymphocytes that leads to their compartmentalized distribution. *The T cell zone* contains a denser reticular meshwork and produces CCL19 and CCL21 chemokines attracting T cells (Luther et al. 2000). *The follicles* are more heterogeneous with regard to their reticular framework, as their nonhemopoietic compartment also includes follicular dendritic cells (FDCs) and possibly other FRC subsets with distinct cell surface phenotype (Link et al. 2007). FDCs are a dominant source for CXCL13 chemokine that recruits B cells and a small CD4 T cell subset into the follicles (Ansell et al. 2000). In addition to the fibroblastic reticular meshwork cosegregating with the main lymphoid compartments into T-zone and follicles, specialized zones formed by FRCs with distinct phenotypic and functional attributes have been described at the *periphery of B cell follicles* in both spleen and lymph nodes of mice (Balogh et al. 2004; Katakai et al. 2008). This compartment – variably referred to as circumferential reticulum in spleen or marginal reticular cells (MRCs) in lymph nodes – encloses the white pulp and also serves as a stromal support for the marginal sinus or perifollicular space. A fibroblastic stromal microdomain, which may correspond to such territory in human spleen, may be the perifollicular zone which, similarly to the mouse counterpart, contains fibroblasts expressing MAdCAM-1 glycoprotein (Steiniger et al. 2001). As MAdCAM-1 can also be expressed by a variety of stromal cells (undifferentiated stromal mesenchyme, endothelium, fibroblasts/marginal reticular cells, follicular dendritic cells), the lineage affiliation and differentiation stage of these MAdCAM-1 positive cells in either human or mouse spleen are uncertain. In rodents, the MZ does not enclose completely the white pulp, but some gaps (called MZ bridging channels) remain where the PALS is placed in the immediate vicinity of red pulp (Mitchell 1973; Bajenoff et al. 2008). At these locations, a direct contact is permitted between the FRC-rich T-zone compartment and the marginal sinus, which is used as guidance by recirculating lymphocytes to reach their follicular or PALS destination within the white pulp after extravasation.

In contrast to the elaborate structural organization of the white pulp and MZ, the architecture of *splenic red pulp* is less well defined. Its main elements are red pulp cords (of Billroth), complex structures formed by fibroblastic cells, macrophages, and, to a lesser extent, lymphocytes, including plasma cells. The positioning of plasma cells is promoted by the chemokine production (CXCL12) of stromal cells as well as their integrin-mediated retention by ICAM1–LFA1 interaction (Hargreaves et al. 2001; Ellyard et al. 2005), whereas the red pulp macrophages are mostly resident hemopoietic cells. These red pulp cords also enclose venous sinusoids and terminal segments of penicilliform arterioles, which vascular beds represent the open circulation of the splenic vasculature. The basement membrane of venous sinusoids is fenestrated and the endothelial cells are only loosely connected, where interendothelial slits facilitate the extravasation of blood cells.

The venous capillaries drain into larger collecting veins, which join to form the trabecular veins. Passage through the sinus wall requires a certain deformability of the reentering blood cells, thus the cells with increased fragility will be retained within the splenic cords and removed by local macrophages.

More recently, the *conduit system*, a novel structural entity involved in the fluid transport in the white pulp of the spleen, has been described (Nolte et al. 2003). Originally discovered in the T cell zone of lymph nodes (Gretz et al. 1996), these structures are formed by the unique manner how FRCs typical for peripheral lymphoid organs and thymus are arranged. Instead of being embedded in extracellular matrix components, they attach to a set of tubular structures enclosed by basement membrane. The center of the tubules is built up from collagen fibres, which are stabilized both to each other and the enwrapping basement membrane underneath the FRCs by a variety of cross-linking molecules (Lokmic et al. 2008). The presence of these structures in lymphoid organs simultaneously separates the "fluid" phase of extravascular spaces from the static "parenchymal" (i.e. lymphocytic) territories and also limits the size of accessible compounds to approximately 70 kDa. This size limit appears somewhat larger than in lymph nodes, where smaller molecules are permitted to enter the conduits, whereas larger molecules and physical particles are removed by macrophages located underneath the planar floor of subcapsular sinus. In the spleen, the diameter of these tubules shows a regional variation, where the widest conduits are in the PALS and, to a lesser extent, in the follicles, which are followed by the MZ, and the narrowest conduits can be found in the red pulp area. The presumed function of these conduits is the display of homeostatic chemokines for the directional movement of lymphocytes (see Sect. 12.2); however, how lymphocytes may contact with these "shielded" molecules on the basement membrane aspect of reticular cells (therefore at the less accessible aspect of the cell) is currently debated. At regions with gaps of FRCs envelope, dendritic cells appear to occupy these sites, offering the possibility for sampling the content of conduit (Bajénoff et al. 2006). As most of these observations were obtained in mouse, it remains to be seen whether human spleens also possess such structures with similar functions.

## 11.3 Transcriptional Regulation of Early Spleen Development

The development of spleen in mouse can be divided into three main phases. The first phase begins with the condensation of mesenchymal cells in the dorsal mesogastrium along its left side, leading to the formation of splenic anlage dorsal to the stomach. This region is adjacent to the splanchnic mesodermal plate (SMP), an initially symmetric layer of columnar mesenchymal cells, which may act as an important early organizer of spleen development (Hecksher-Sørensen et al. 2004). Subsequently, hemopoietic cells colonize the spleen, coupled with an early segregation of red pulp and rudimentary periarteriolar white pulp regions in the second phase. Third, mature lymphoid cells (initially B cells at around the birth, followed by T cells during the first postnatal week) accumulate gradually in periarteriolar location. As a result, the follicular architecture is established, including the differentiation of FDCs and compartmentalization of FRC meshwork. Finally, the MZ will emerge during the next 2–3 postnatal weeks. Thus in mouse spleen, a considerable part of organ development is delayed until the perinatal and early postnatal period. In human subjects, however, a more accelerated developmental kinetics dictates an earlier formation of spleen, resulting in differentiated follicles containing FDCs and extensive T cell zones with supporting fibroblastic meshwork approximately as early as around the first half of second trimester (Steiniger et al. 2007). On the other hand, relatively little is known about the first phase of human spleen development.

The detailed analysis of the effects of spontaneous or directed mutations of several genes encoding morphogenic factors in mice proved invaluable for understanding the genetic regulation of early developmental events of spleen. Despite the significant advance in mapping several transcription factors to developing spleen in mice (Brendolan et al. 2007), however, both their functions and hierarchic relationship in controlling the spleen formation are still unknown.

The first mutation with spleen defect was the dominant hemimelia (Dh), where the identity of affected gene(s) is unknown to date. Together with limb abnormalities, this spontaneous autosomal dominant mutation also results in several visceral defects, including the absence of spleen in both heterozygotes and homozygote mutants (Green 1967). The effect of Dh probably causes the absence of SMP (Hecksher-Sørensen et al. 2004).

Subsequent to its symmetric appearance, SMP is induced to undergo a leftward growth, while the right-sided component is lost in less than 2 days. This asymmetric growth is influenced by Pitx2 and Barx1 transcription factors participating in the left–right signaling cascade (Patterson et al. 2000; Kim et al. 2007), and the process possibly involves fibroblast growth factors 9 and 10 (FGF9 and FGF10) under the control of homeodomain-type transcription factor Bapx1/Nkx3.2 (Hecksher-Sørensen et al. 2004), whose absence results in asplenia (Lettice et al. 1999). This process may also be regulated by Sox11, a member of the mammalian group C Sox factor family (Sock et al. 2004). In addition to the leftward growth, the splenic anlage also approaches the anterior part of the stomach in a process requiring a continuous display of yet unidentified directional cues (Burn et al. 2008). In Bapx1-deficient mice, the separation between pancreas and spleen does not occur. Instead the pancreatic endoderm and the surrounding splenic mesenchyma undergo a metaplastic transformation to form gut-like structure in a process which also involves signaling along the Sonic hedgehog (Shh) pathway (see below).

Bapx1 is also involved in the upregulation of several genes required for spleen development. These include Hox11/Tlx1, which was the first targeted homeobox gene demonstrated necessary for spleen development (Roberts et al. 1994). In normal spleen development, both the slightly delayed appearance of Tlx1 relative to that of Bapx1 and its absence in Bapx1-deficient mice suggest that Hox11/Tlx1 in the regulatory hierarchy is subordinated to Bapx1 in the mesenchyma. On the other hand, Tlx1 was necessary for the expression of Wt1 (Wilms tumor suppressor gene), whose absence also causes asplenia, presumably due to proliferation defects during the later phase of spleen development (Dear et al. 1995; Koehler et al. 2000). The absence of Bapx1 does not affect the expression of Nkx2–5, a spleen-specific homeodomain transcription factor, which also acts as a master regulator of cardiac development (Brendolan et al. 2005).

More recently, another master regulator for spleen development was also identified, whose regulatory pathway may partially overlap with that of Bapx1. In mice with inactivated Pbx1 homeobox transcription factor, the SMP is preserved (unlike in Dh mutation with absent SMP), but the spleen fails to form subsequently. Pbx1 is a member of the TALE (three-amino acid loop extension) homeodomain family, and it is required for the formation of several visceral organs. The lack of Pbx1 also causes the loss of Tlx1 and Wt10 and, in addition, the absence of Nkx2–5, another early spleen-associated transcription factor. The absence of Pbx1, however, does not influence the expression of Bapx1. It appears, therefore, that Tlx1 and Wt1 are independently regulated by Bapx1 and Pbx1, and both pathways are simultaneously needed for the proper spleen development. Furthermore, the Pbx1-dependent expression of Wt1 involving Tlx1 as intermediate is probably focused onto the condensing internal mesenchymal part of the spleen and not the external part which will form the capsule (Brendolan et al. 2005).

In addition to the Bapx1 and Pbx1 as upstream regulators which both upregulate Wt1 through Tlx1, the absence of another transcription factor Pod1/capsulin/Tcf21 also results in asplenia (Lu et al. 2000). Pod1 is a member of bHLH (basic helix-loop-helix) transcription factor family. The members of this family form homo- or heterodimeric complexes for target binding. The regulatory process in which Pod1/Tcf21 exerts its effect involves Nkx2–5 as one potential target gene. Similarly to Pbx1, in the absence of Pod1, the expression of Nkx2–5 is blocked, whereas its effect on the Tlx1-mediated upregulation of Wt1 is unknown. Thus in the early spleen specification three parallel, yet independent and nonredundant, pathways have been proposed to be involved (Brendolan et al. 2005). Of these, Bapx1 effect is mediated via Tlx1-Wt1 pathway, without the involvement of Nkx2–5. According to the proposed scheme, Pbx1 has dual activity, one involving the Tlx1/Wt1 pathway, the other affects Nkx2–5. Similarly to this latter connection, Pod1 also influences spleen development by upregulating Nkx2–5.

The expression of gene Hox11/Tlx homeodomain transcription factor at embryonic day E11,5 indicates the genetic commitment of the splenic subserosal mesenchyma in mouse embryos to differentiate towards the spleen fate. This process is further influenced by interactions between endodermal and mesodermal layers of gut, which involves signaling via soluble Sonic hedgehog (Shh) protein produced by the lateral endoderm, and its receptor Patched (Ptch1) and associating components. Upon binding, Ptch1 is relieved from repressing of Smoothened (Smo), a G protein-coupled receptor-like protein. As a result, the activity of several transcription factors is altered as the activator forms(s) of GLI protein as mediator targets downstream genes with the appropriate recognition sequence (Jenkins 2009). The lack of expression of both Shh and Indian hedgehog (Ihh) at the antero-posterior segments of gut is required to maintain the pancraticosplenic potential of mesenchyma in posterior foregut (Apelqvist et al. 1997). This restriction of Shh signaling is probably maintained by Activin receptor ActRII engagement by various ligands, including activins (Kim et al. 2000). ActRIIA and ActRIIB may bind activins, members of TGFβ superfamily and use Smad2 and Smad3 for signaling complexed with Smad4 (Abe et al. 2004). In addition to modulating Shh signaling, ActRIIB is also instrumental in positioning the spleen to the left side of gut tube by binding *nodal* ligand (Oh and Li 2002). The exact relationship between the Shh signaling, activin receptor functions, FGF8 as soluble factor, and Bapx1 as potential target gene for left–right asymmetry (Schneider et al. 1999) is still unresolved.

Subsequent to the splenic specification of mesenchyma, an extensive vascularization ensues along spleen-specific patterning. This latter process at around embryonic day E15,5 involves the activity of Nkx2–3 homeodomain transcription factor (Pabst et al. 1999; Wang et al. 2000). Its absence manifests in a complex vascular malformation of both the red pulp and marginal sinus, in addition to defective MZ development and aberrant white pulp FRC meshwork (Balogh et al. 2007; Bovári et al. 2007). Nkx2–3 transactivates the promoter of MAdCAM-1 adhesion protein, which may contribute to the structural defects of the splenic MZ (Pabst et al. 2000). In addition to the splenic defects, the absence of Nkx2–3 also affects the formation of Peyer's patches, although the peripheral lymph nodes are normal (Wang et al. 2000). A recent implication of the involvement of Nkx2–3 in pathological intestinal lymphoid tissue development may be the association of SNP variants of Nkx2–3 with both Crohn's disease and ulcerative colitis in humans (Cho 2008).

## 11.4 Emergence of the Spleen Anlage and Specification of Red Pulp and White Pulp

As a result of splenic specification and commitment, the splenic anlage rapidly separates from the stomach and dorsal pancreas and becomes a hemopoietic tissue, with no discernible organizational pattern.

The maturation of splenic tissue includes the high-degree vascularization of spleen, where vessels with different endothelial cell surface phenotype can be observed after the separation of the organ from the stomach. The embryonic arteries display VE-cadherin, CD31, and Tie-2 as endothelial restricted markers, but are negative for MECA-32 antigen, while the smaller vessels coexpressed VE-cadherin and MECA-32 marker. In addition, these developing vessels also display MAdCAM-1 and ICAM-1 adhesion molecules, and the arterioles are surrounded by VCAM-1$^{+}$ reticular cells (Vondenhoff et al. 2008). Arteriolar endothelial cells produce CCL19, CCL21, CXCL12, and CXCL13 chemokines, whereas perivascular stromal cells produce only the latter two compounds. Possibly as a result of the concerted action of these chemoattractants, at the peripheral segment of perivascular mesenchyma lymphoid cells displaying lymphoid tissue inducer phenotype

(LTi, $CD4^{+}/CD3^{-}/IL\text{-}7R\alpha^{+}/ROR\gamma t^{+}$) appear E13,5 onward, although unlike their counterparts in lymph nodes, their presence is not required for the development of spleen. On the other hand, their presence likely generates reverse signals for the perivascular stroma to upregulate VCAM-1 expression, and the subsequent colonization of T cells to the LTi-clustered region also possibly involves lymphotoxin-dependent signals in the early postnatal period (Withers et al. 2007). In addition, the fetal spleen stromal microenvironment also promotes the expansion of myeloid cells with the phenotype of macrophages found postnatally in the red pulp (Bertrand et al. 2006).

In humans, the available information concerning the early organogenesis of splenic anlage is sparse, and most of these are based on morphological analyses including ultrastructural studies and immunohistological observations. Virtually, no data are available concerning the expression of fate-determining transcription factors of primordial spleen that have been extensively studied in mice (see above). Furthermore, in contrast to rodent spleen, in human embryos the erythromyelopoietic (Calhoun et al. 1996) or B-lymphopoietic (Asma et al. 1984) activity of spleen is either minimal or absent, although the spleen contains a large number of immature hemopoietic cells, with a substantial fraction being in circulation.

According to morphological characteristics of vasculature and degree of lymphoid compartmentalization, the subsequent embryonic and early neonatal formation of spleen can be divided into four stages (Steiniger et al. 2007a). The formation of human spleen at 35–40 days begins after ovulation (corresponding to stage 0) in the left-dorsal part of the mesogastrium, initiated by the cluster formation of reticular cells at around the 8th week after ovulation, which is colonized by macrophages (Ishikawa 1985). In phase I, the spleen is segmented into arterial lobules (Vellguth et al. 1985). At this stage, evenly scattered lymphocytes could be observed, while CXCL13 chemokine is already produced. The lymphoid cells are mostly B cells that form periarteriolar cluster. Subsequently in stage II between 18 and 30 weeks, the lobular structure gradually dissolves by the enrichment of venous sinus network originating from the venules at the peripheral parts of the former lobules, while the lymphoid accumulation becomes more complex: T cells preferentially accumulate at the more central regions of arteriolar segments into a reticular meshwork formed by mesenchymal cells expressing α-smooth muscle actin (αSMA) that also produce CCL21 chemokine, whereas the B cells are located more peripherally. These αSMA-positive cells also outline the developing MZ at around the 26th week of pregnancy (Satoh et al. 2009). In parallel, the reticular-stromal mesenchymal also becomes more heterogeneous, including the formation of FDCs in areas with the paucity of αSMA-positive reticular cells. In contrast to the previous perivascular accumulation of B cells into nonfollicular arrangement, the formation of follicles takes place at around larger caliber arterioles in the vicinity of T-zone, and with the simultaneous differentiation of FDCs. At stage III, the lobular structure disappears, while the follicles are developed, although perivascular nonfollicular B cells are also detectable The formation of macrophage sheaths with sialoadhesin/$CD169^{+}$ macrophages probably takes place after birth, where some subtle alterations of red pulp vasculature (appearance of various red pulp sinus subsets with differential expression of CD141 and CD8) are also likely to occur (Steiniger et al. 2007b).

## 11.5 Specialization of the Mesenchymal Scaffolding of the Mouse Spleen: Postnatal Evolution of Vasculature and Fibroblastic Reticular System

The intrauterine phase of spleen formation in mouse is sufficient to complete its developmental programming to reach the stage of a separate organ, compartmentalized into white and red pulp. However, its architecture is still immature and needs further structural rearrangements to occur in order to enable it to engage in adaptive immune responses. These alterations include the emergence of compartmentalized fibroblastic architecture of white pulp, differentiation of follicular stroma, and the establishment of MZ, coupled with the colonization of these compartments with mature lymphocytes

Immediately after birth, the white pulp territory is poorly organized around small-to-medium sized arterioles, which are surrounded by FRCs expressing VCAM-1 and MAdCAM-1 and smooth muscle cells (Withers et al. 2007). Subsequent outward migration of FRCs towards the edge of the white pulp is coupled with the gradual accumulation of B cells (including a relatively large number of $CD5^{+}$/B-1 cells) and T cells (Wen et al. 2005; Balogh et al. 2004). Their presence supplies those TNF/lymphotoxin signals that are required for the induction of follicular stroma, particularly FDCs (De Togni et al. 1994; Pasparakis et al. 1996; Matsumoto et al. 1996; Alimzhanov et al. 1997; Koni et al. 1997). The accumulation of B cells and subsequent differentiation of FDCs together with the stabilization of follicles appear to be dictated by two signaling events. In early period, an initial stimulus is mediated by LTβR and associating molecules of the noncanonical NF-κB NIK-RelB pathway, without the canonical pathway with TNFR-associated factor 6 (TRAF6) as adaptor; however, the stabilization during the second postnatal week requires both the continuation of LTβR engagement and the activation of TRAF6-mediated signaling (Qin et al. 2007). The immature precursors of FDCs initially cluster at the edge of white pulp in day 3 old newborn mice and, possibly upon the engagement their TNFR-I, migrate into the B cell rich follicles during the first week of postnatal period (Pasparakis et al. 2000). Thus, the differentiation of follicular stromal cells is associated both with their functional specification, as manifested by their gradual acquisition of function-related markers including complement receptor and CXCL13 secretion, and also their physical relocation from the periphery to the inner part of developing follicle. The recent identification of a commonly used marker for the detection of mouse FDCs (FDC-M1) expressed even in immature stage of FDCs (Balogh et al. 2001) as milk fat globule epidermal growth factor 8 (Mfge8), a known phosphatidylserine binding compound and ligand for integrin-mediated recognition (Kranich et al. 2008), will facilitate the fate mapping of these cells.

The postnatal accumulation of B cells in the developing spleen is also necessary for the increased production of CCL19 and CCL21 chemokines by regional FRCs attracting T cells to the developing T-zone (Ngo et al. 2001). Similarly to FDC-repositioning, in mice a gradual rearrangement of FRC subsets can be observed during the first two postnatal weeks, resulting in their accumulation within the PALS and in a shell encircling the follicles formed underneath the MZ that develops simultaneously (Balogh et al. 2004). The characteristic phenotypic markers attributed to murine FRC subsets are listed in Table 11.1.

Although the presence of B cells has been demonstrated to be important for the establishment of white pulp stromal architecture and MZ, it is uncertain whether B cells at which stage of differentiation (immature or mature) or lineage affiliation (follicular or MZ descendants within the B-2 lineage or B-1 subset) are capable of providing the necessary signals for stromal maturation. Studies on the postnatal formation of FDCs have revealed that splenic FDC precursors at various tissue locations are detectable as early as 5–7 postnatal day, when the overwhelming majority of B cells are still immature, and despite the radiation-induced depletion of lymphocytes, the differentiation of immature FDCs continues (Balogh et al. 2001). In addition, at this early stage, the dominant fraction of splenic B cells belongs to the B-1 lineage (Wen et al. 2005), and their high CXCR5 expression predisposes these cells to contact with uncommitted FDC-precursors producing low-amount of CXCL13 over the B-2 lineage (and still immature) B cells with lower level of CXCR5 display. Thus, it may be worthwhile to repeat the previous philosophical question on the biological meaning for LTi cells to induce lymph node formation: what kind of "intuition" can be expected to be possessed by (fetal liver-derived)

**Table 11.1** Secreted or membrane-associated markers used to identify murine FRCs and their subsets

| Marker | Molecular characteristics and functions | Expression in the reticular meshwork | LT-dependence |
|---|---|---|---|
| ER-TR7 | Conduit material | White pulp, marginal zone, red pulp | No |
| BP-3 | GPI-linked antigen related to CD38 | Mainly follicle-associated FRCs, T/B border region | Yes |
| Podoplanin/gp38 | Mucin-type glycoprotein with unknown function | T-zone FRCs, T/B border region | Yes |
| MAdCAM-1 | Peyer's patches and mesenteric lymph node homing addressin | Marginal zone and circumferential reticulum | Yes |
| L1/NCAM | Ig-superfamily member; cell adhesion molecule | Marginal zone and circumferential reticulum | Unknown |
| IBL-10 | Unknown | Red pulp trabecules, follicles and circumferential reticulum, T/B border, T-zone FRCs | Unknown |
| IBL-11 | Unknown | Circumferential reticulum, T-zone FRCs | Unknown |
| CCL19, CCL21 | Homeostatic chemokines | T-zone FRCs | Yes |

neonatal B-1 lymphocytes or other immature B-lineage cells for initiating the formation of a lymphoid tissue territory (follicles) which will be colonized by bone marrow-derived B cells, and from which the "founder" B-1 lymphocytes later will be largely excluded?

Immediately after birth, the splenic vessels in mouse lack any noticeable organizational pattern, although the red pulp area is already richly perfused with sinusoids, and B cell accumulation may surround arterioles in the primordial white pulp. The enlargement of spleen needs to be supported by the extension of its vasculature by the formation of new vessels. Crucial angiogenic factors such as Ang-1, Ang-2, and VEGF may play a role, as their receptors (Tie-2 and VEGFR-2, respectively) are present on neonatal splenic endothelium (Zindl et al. 2009). In the next few days, a gradual endothelial clustering can be observed. By the end of the first week, a marginal sinus with loosely organized endothelial cells is formed. The migration of putative marginal sinus forming cells is probably controlled by lymphotoxin and TNF signaling, where the stimulatory signals are delivered during the first few postnatal days. It is not yet known what chemotactic and adhesive signals regulate the regional clustering of marginal sinus endothelium cells, although some role of MAdCAM-1 addressin in organizing the endothelial clustering is likely. In addition, some components of the marginal sinus-lining cells may derive from other lineage, as reporter genes driven by endothel-specific promoters such as eNOS or vWF, failed to indicate regional expression (Aird et al. 1995).

**Table 11.2** Membrane antigens used to identify endothelial cells and their subsets in the mouse spleen

| Marker | Molecular characteristics and functions | Expression in the splenic vasculature | LT-dependence |
|---|---|---|---|
| CD31 PECAM-1 | ITIM-bearing member of Ig-superfamily involved in homophilic and heterophilic binding (CD38 and heparin-dependent proteoglycans) | Central arteriole (strongly), white pulp arteriolar segments, marginal sinus and red pulp sinus subset (weakly) | No |
| MAdCAM-1 | Peyer's patches and mesenteric lymph node homing addressin | Marginal sinus-lining cells | Yes |
| VE-cadherin CD144 | Ca-dependent homophilic adhesion of vascular endothelial cells | Central arteriole, white pulp arteriolar segments, marginal sinus and red pulp sinuses | No |
| VEGFR-2/flk1 | Receptor for VEGF | Central arteriole, white pulp arteriolar segments, marginal sinus and red pulp sinus subset | Yes* |
| Stabilin-2 | Hyaluronane receptor | Red pulp sinuses | Unknown |
| IBL-7/1 | Unknown (distinct from VEGFR-2) | White pulp arteriolar segments, marginal sinus and red pulp sinus subset | Yes* |
| IBL-9/2 | Unknown | Red pulp sinus subset | No |

*Distributional differences observed in mice with LT signaling defects

Furthermore, other components, such as MZ macrophages, may also be instrumental for structural integrity, as their perturbation leads to a delayed formation of marginal sinus (Chen et al. 2005). In the absence of B cells, the formation of marginal sinus is severely impaired, and the maintenance of the complex MZ also requires B cells (Nolte et al. 2004). However, the cellular signals for the structural integrity of MZ do not necessarily derive from marginal zone B (MzB) cells, as Sn/CD169$^{+}$ macrophages with normal distribution are also present in mice lacking MzB lymphocytes due to the absence of tyrosine kinase Pyk-2 or its upstream regulator Rap1b (Guinamard et al. 2000; Chen et al. 2008). The topographic analysis of postnatal development of splenic vasculature relies on the availability of reagents identifying distinct endothelial subsets in the adult spleen. Table11.2-summarizes the available antibodies recognizing endothelium-associated markers in mice and their reactivity against various subsets. Figure 11.2 illustrates the architecture of splenic vasculature in mice.

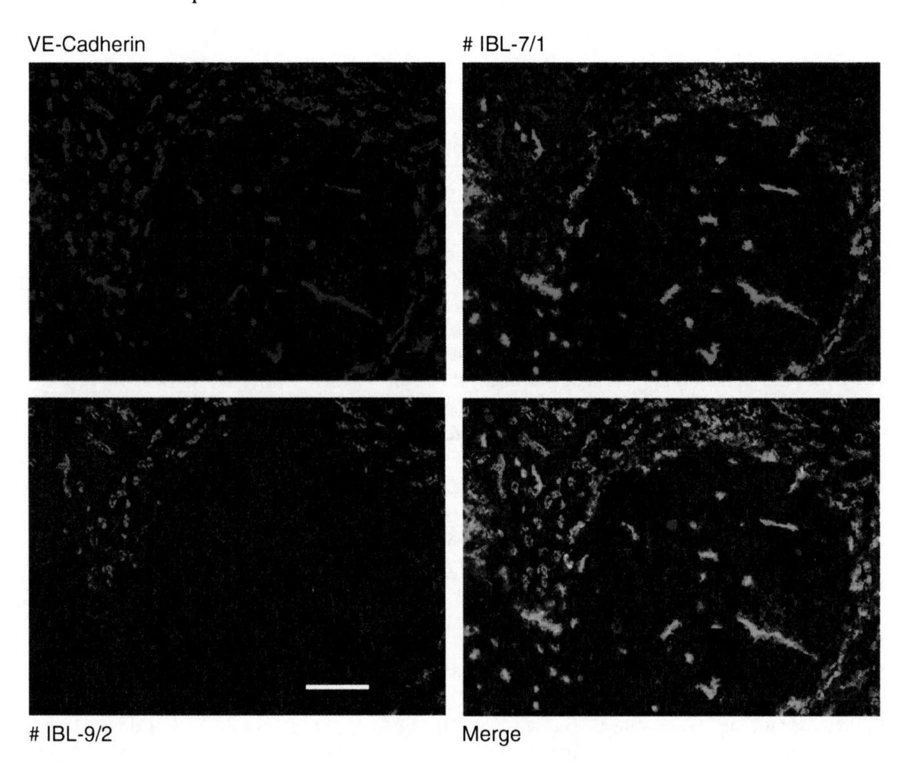


**Fig. 11.2** In addition to its highly compartmentalized hemopoietic (lymphoid as well as macrophage) distribution, the endothelia lining separate vascular beds of spleen display a remarkable phenotypic heterogeneity. Thus endothelial cells identified by VE-cadherin (CD144) pan-endothelial marker (in *blue*) line vascular segments in the red pulp (RP), marginal zone (MZ), and white pulp (WP). In contrast, IBL-9/2 mAb (in *red*) highlights only RP vessels, which also variably express IBL-7/1 (*green*) antigen, most accentuated in the marginal sinus, in addition to some scattered sinuses in the RP and terminal arterioles in the WP. The endothelial cells in the IBL-7/1$^{+}$ segments of MZ and adjacent fibroblastic cells also display MAdCAM-1 (not shown). Scale bar = 250 μm

In addition to establishing an important filtering compartment for blood-borne pathogens, the MZ is also crucially involved in the splenic homing of recirculating lymphocytes. In lymph nodes, the homing process begins with the selective binding of leukocytes to the luminal surface of high endothelial venules, mediated by CD62L and PNAd molecules. In spleen, the identity of corresponding endothelial or other – probably MZ macrophage – partner is unknown, as are the participants of possible ligand–counterreceptor pairs (Nolte et al. 2002). Furthermore, considering the complex process of postnatal MZ formation together with the gradual development of marginal sinus in mouse, it is also conceivable that the characteristics of splenic homing process (kinetics, cellular partners, and molecular participants) may also change, reflecting these alterations.

**Acknowledgment** This work was supported by the University of Pécs Faculty of Medicine Research Grant to Péter Balogh.

## References

Abe Y, Minegishi T, Leung PC (2004) Activin receptor signaling. Growth Factors 22:105–110

Agius C, Roberts RJ (2003) Melano-macrophage centres and their role in fish pathology. J Fish Dis 26:499–509

Aird WC, Jahroudi N, Weiler-Guettler H, Rayburn HB, Rosenberg RD (1995) Human von Willebrand factor gene sequences target expression to a subpopulation of endothelial cells in transgenic mice. Proc Natl Acad Sci USA 92:4567–4571

Alimzhanov MB, Kuprash DV, Kosco-Vilbois MH, Luz A, Turetskaya RL, Tarakhovsky A, Rajewsky K, Nedospasov SA, Pfeffer K (1997) normal development of secondary lymphoid tissues in lymphotoxin beta-deficient mice. Proc Natl Acad Sci USA 94:9302–9307

Ansell K, Ngo VN, Hyman PL, Luther SA, Förster R, Sedgwick JD, Browning JL, Lipp M, Cyster JG (2000) A chemokine-driven positive feedback organizes lymphoid follicles. Nature 406:309–314

Apelqvist A, Ahlgren U, Edlund H (1997) Sonic hedgehog directs specialized mesoderm differentiation in the intestine and pancreas. Curr Biol 7:801–804

Asma GE, Langlois van den Bergh R, Vossen JM (1984) Development of pre-B and B lymphocytes in the human fetus. Clin Exp Immunol 56:407–414

Bajénoff M, Egen JG, Koo LY, Laugier JP, Brau F, Glaichenhaus N, Germain RN (2006) Stromal cell networks regulate lymphoid entry, migration and territoriality in lymph nodes. Immunity 25:989–1001

Bajenoff M, Glaichenhaus N, Germain RN (2008) Fibroblastic reticular cells guide T lymphocyte entry into and migration within the splenic T cell zone. J Immunol 181:3947–3954

Balázs M, Horváth G, Grama L, Balogh P (2001) Phenotypic identification and development of distinct microvascular compartments in the postnatal mose spleen. Cell Immunol 212: 126–137

Balogh P, Aydar Y, Tew JG, Szakal AK (2001) Ontogeny of the follicular dendritic cell phenotype and function in the postnatal murine spleen. Cell Immunol 214:45–53

Balogh P, Horváth G, Szakal AK (2004) Immunoarchitecture of distinct reticular fibroblastic domains in the white pulp of mouse spleen. J Histochem Cytochem 52:1287–1298

Balogh P, Balázs M, Czömpöly T, Weih DS, Arnold HH, Weih F (2007) Distinct roles of lymphotoxin-beta signaling and the homeodomain transcription factor Nkx2.3 in the ontogeny of endothelial compartments in spleen. Cell Tissue Res 328:473–486

Bertrand JY, Desanti GE, Lo-Man R, Leclerc C, Cumano A, Golub R (2006) Fetal spleen stroma drives macrophage commitment. Development 133:3619–3628
Bovári J, Czömpöly T, Olasz K, Arnold HH, Balogh P (2007) Complex organizational defects of fibroblast architecture in the mouse spleen with Nkx2.3 homeodomain deficiency. Pathol Oncol Res 13:227–235
Brendolan A, Ferretti E, Salsi V, Moses K, Quaggin S, Blasi F, Cleary ML, Selleri L (2005) A Pbx1-dependent genetic and transcriptional network regulates spleen ontogeny. Development 132:3113–3126
Brendolan A, Rosado MM, Carsetti R, Selleri L, Dear TN (2007) Development and function of the mammalian spleen. Bioessays 29:166–177
Burn SF, Boot MJ, de Angelis C, Doohan R, Arques CG, Torres M, Hill RE (2008) The dynamics of spleen morphogenesis. Dev Biol 318:303–311
Calhoun DA, Li Y, Braylan RC, Christensen RD (1996) Assessment of the contribution of the spleen to granulocytopoiesis and erythropoiesis of the mid-gestation human fetus. Early Hum Dev 46:217–227
Chen Y, Pikkarainen T, Elomaa O, Soininen R, Kodama T, Kraal G, Tryggvason K (2005) Defective microarchitecture of the spleen marginal zone and impaired response to a thymus-independent type 2 antigen in mice lacking scavenger receptors MARCO and SR-A. J Immunol 175:8173–8180
Chen Y, Yu M, Podd A, Wen R, Chrzanowska-Wodnicka M, White GC, Wang D (2008) A critical role of Rap1b in B-cell trafficking and marginal zone B-cell development. Blood 111:4627–4636
Cho JH (2008) The genetics and immunopathogenesis of inflammatory bowel disease. Nat Rev Immunol 8:458–466
Dear TN, Colledge WH, Carlton MB, Lavenir I, Larson T, Smith AJ, Warren AJ, Evans MJ, Sofroniew MV, Rabbitts TH (1995) The Hox11 gene is essential for cell survival during spleen development. Development 121:2909–2915
De Togni P, Goellner J, Ruddle NH, Streeter PR, Fick A, Mariathasan S, Smith SC, Carlson R, Shornick LP, Strauss-Schoenberger J, Chaplin DD (1994) Abnormal development of peripheral lymphoid organs in mice deficient in lymphotoxin. Science 264:703–707
Del Cacho E, Gallego M, Sanz A, Zapata A (1993) Characterization of distal lymphoid nodules in the chicken caecum. Anat Rec 237:512–517
Del Cacho E, Gallego M, Arnal C, Bascuas JA (1995) Localization of splenic cells with antigen-transporting capability in the chicken. Anat Rec 241:105–112
Du Pasquier L, Robert J, Courtet M, Mussmann R (2000) B-cell development in the amphibian Xenopus. Immunol Rev 175:201–213
Ellyard JI, Avery DT, Mackay CR, Tangye SG (2005) Contribution of stromal cells to the migration, function and retention of plasma cells in human spleen: potential roles of CXCL12, IL-6 and CD54. Eur J Immunol 35:699–708
Fänge R, Nilsson S (1985) The fish spleen: structure and function. Experientia 41:152–158
Flajnik MF (2002) Comparative analyses of immunoglobulin genes: surprises and portents. Nat Rev Immunol 2:688–698
Green MC (1967) A defect of the splanchnic mesoderm caused by the mutant gene dominant hemimelia in the mouse. Dev Biol 15:62–89
Gretz JE, Kaldjian EP, Anderson AO, Shaw S (1996) Sophisticated strategies for information encounter in the lymph node: the reticular network as a conduit of soluble information and a highway for cell traffic. J Immunol 157:495–499
Guinamard R, Okigaki M, Schlessinger J, Ravetch JV (2000) Absence of marginal zone B cells in Pyk-2-deficient mice defines their role in the humoral response. Nat Immunol 1:31–36
Hansen JD, Zapata AG (1998) Lymphocyte development in fish and amphibians. Immunol Rev 166:199–220
Hargreaves DC, Hyman PL, Lu TT, Ngo VN, Bidgol A, Suzuki G, Zou YR, Littman DR, Cyster JG. (2001) A coordinated change in in chemokine responsiveness quides plasma cell movements. J Exp Med 194:45–56

Hecksher-Sørensen J, Watson RP, Lettice LA, Serup P, Eley L, De Angelis C, Ahlgren U, Hill RE (2004) The splanchnic mesodermal plate directs spleen and pancreatic laterality, and is regulated by Bapx1/Nkx3.2. Development 131:4665–4675

Igyártó BZ, Magyar A, Oláh I (2007) Origin of follicular dendritic cell in the chicken spleen. Cell Tissue Res 327:83–92

Ishikawa H (1985) Differentiation of red pulp and evaluation of hemopoietic role of human prenatal spleen. Arch Histol Jpn 48:183–197

Ito S, Naito M, Kobayashi Y, Takatsuka H, Jiang S, Usuda H, Umezu H, Hasegawa G, Arakawa M, Shultz LD, Elomaa O, Tryggvason K (1999) Roles of a macrophage receptor with collagenous structure (MARCO) in host defense and heterogeneity of splenic marginal zone macrophages. Arch Histol Cytol 62:83–95

Jenkins D (2009) Hedgehog signalling: emerging evidence for non-canonical pathways. Cell Signal 21:1023–1034

Jeurissen SHM (1993) The role of various compartments in the chicken spleen during an antigen-specific humoral response. Immunology 80:29–33

Kaiser P, Poh TY, Rothwell L, Avery S, Balu S, Pathania US, Hughes S, Goodchild M, Morrell S, Watson M, Bumstead N, Kaufman J, Young JR (2005) A genomic analysis of chicken cytokines and chemokines. J Interferon Cytokine Res 25:467–484

Katakai T, Suto H, Sugai M, Gonda H, Togawa A, Suematsu S, Ebisuno Y, Katagiri K, Kinashi T, Shimizu A (2008) Organizer-like reticular stromal cell layer common to adult secondary lymphoid organs. J Immunol 181:6189–6200

Kim BM, Miletich I, Mao J, McMahon AP, Sharpe PA, Shivdasani RA (2007) Independent functions and mechanisms of homeobox gene Barx1 in patterning mouse stomach and spleen. Development 134:3603–3613

Kim SK, Hebrok M, Li E, Oh SP, Schrewe H, Harmon EB, Lee JS, Melton DA (2000) Activin receptor patterning of foregut organogenesis. Genes Dev 14:1866–1871

Koehler K, Franz T, Dear TN (2000) Hox11 is required to maintain normal Wt1 mRNA levels int he developing spleen. Dev Dyn 218:201–206

Koni PA, Sacca R, Lawton P, Browning JL, Ruddle NH, Flavell RA (1997) Distinct roles in lymphoid organogenesis for lymphotoxins alpha and beta revealed in lymphotoxin beta-deficient mice. Immunity 6:491–500

Kraal G, Schornagel K, Streeter PR, Holzmann B, Butcher EC (1995) Expression of the mucosal vascular addressin, MAdCAM-1, on sinus-lining cells in the spleen. Am J Pathol 147: 763–771

Kranich J, Krautler NJ, Heinen E, Polymenidou M, Bridel C, Schildknecht A, Huber C, Kosco-Vilbois MH, Zinkernagel R, Miele G, Aguzzi A (2008) Follicular dendritic cells control engulfment of apoptotic bodies by secreting Mfge8. J Exp Med 205:1293–1302

Langenau DM, Palomero T, Kanki JP, Ferrando AA, Zhou Y, Zon LI et al (2002) Molecular cloning and developmental expression of Tlx (Hox11) genes in zebrafish (Danio rerio). Mech Dev 117:243–248

Lettice LA, Purdie L A, Carlson GJ, Kilanowski F, Dorin J, Hill RE (1999) The mouse bagpipe gene controls development of axial skeleton, skull, and spleen. Proc Natl Acad Sci USA 96:9695–9700

Link A, Vogt TK, Favre S, Britschgi MR, Acha-Orbea H, Hinz B, Cyster JG, Luther SA (2007) Fibroblastic reticular cells in lymph nodes regulate the homeostasis of naive T cells. Nat Immunol 8:1255–1265

Lokmic Z, Lämmermann T, Sixt M, Cardell S, Hallmann R, Sorokin L (2008) The extracellular matrix of the spleen as a potential organizer of immune cell compartments. Semin Immunol 20:4–13

Lu J, Chang P, Richardson JA, Gan L, Weiler H, Olson EN (2000). The basic helix-loop-helix transcription factor capsulin controls spleen organogenesis. Proc Natl Acad Sci USA 97:9525–9530

Luther SA, Tang HL, Hyman PL, Farr AG, Cyster JG (2000) Coexpression of the chemolines ELC and SLC by T zone stromal cells and deletion of the ELC gene in the plt/plt mouse. Proc Natl Acad Sci USA 97:12694–12699

Martin F, Kearney JF (2002) Marginal zone B cells. Nat Rev Immunol 2:323–335

Marr S, Morales H, Bottaro A, Cooper M, Flajnik M, Robert J (2007) Localization and differential expression of activation-induced cytidine deaminase in the amphibian Xenopus upon antigen stimulation and during early development. J Immunol 179:6783–6789

Matsumoto M, Mariathasan S, Nahm MH, Baranyay F, Peschon JJ, Chaplin DD (1996) Role of lymphotoxin and the type I TNF receptor in the formation of germinal centers. Science 271:1289–1291

Mast J, Goddeeris BM (1999) Development of immunocompetence of broiler chickens. Vet Immunol Immunopathol 70:245–256

Mebius RE, Nolte MA, Kraal G (2004) Development and function of splenic marginal zone. Crit Rev Immunol24:449–464

Mitchell J (1973) Lymphocyte circulation in the spleen Marginal zone bridging channels and their possible role in cell traffic. Immunology 24:93–107

Nagy N, Igyártó B, Magyar A, Gazdag E, Palya V, Oláh I (2005) Oesopphageal tonsil of the chicken. Acta Vet Hung 53:173–188

Nagy N, Oláh I (2007) Pyloric tonsil as a novel gut-associated lymphoepithelial organ of the chicken. J Anat 211:407–411

Ngo VN, Cornall RJ, Cyster JG (2001) Splenic T zone development is B cell dependent. J Exp Med 194:1649–1660

Nolte MA, Hamann A, Kraal G, Mebius RE (2002) The strict regulation of lymphocyte migration to splenic white pulp does not involve common homing receptors. Immunology 106:299–307

Nolte MA, Beliën JA, Schadee-Eestermans I, Jansen W, Unger WW, van Rooijen N, Kraal G, Mebius RE (2003) A conduit system distributes chemokines and small blood-borne molecules through the splenic white pulp. J Exp Med 198:505–512

Nolte MA, Arens R, Kraus M, van Oers MH, Kraal G, van Lier RA, Mebius RE (2004) B cells are crucial for both development and maintenance of the splenic marginal zone. J Immunol 172:3620–3627

Oh SP, Li E (2002) Gene-dosage-sensitive genetic interactions between inversus viscerum (iv), nodal, and activin type IIB receptor (ActRIIB) genes in asymmetrical patterning of the visceral organs along the left-right axis. Dev Dyn 224:279–290

Pabst O, Zweigerdt R, Arnold HH (1999) Targeted disruption of the homeobox transcription factor Nkx2–3 in mice results in postnatal lethality and abnormal development of small intestine and spleen. Development 126:2215–2225

Pabst O, Förster R, Lipp M, Engel H, Arnold HH (2000) NKX2.3 is required for MAdCAM-1 expression and homing of lymphocytes in spleen and mucosa-associated lymphoid tissue. EMBO J 19:2015–2023

Pasparakis M, Alexopoulou L, Episkopou V, Kollias G (1996) Immune and inflammatory responses in TNF-a–deficient mice: a critical requirement for TNF-a in the formation of primary B cell follicles, follicular dendritic cell networks and germinal centers, and in the maturation of the humoral immune response. J Exp Med 184:1397–1411

Pasparakis M, Kousteni S, Peschon J, Kollias G (2000) Tumor necrosis factor and the p55TNF receptor are required for optimal development of the marginal sinus and for migration of follicular dendritic cell precursors into splenic follicles. Cell Immunol 201:33–41

Patterson KD, Drysdale TA, Krieg PA (2000) Embryonic origins of spleen asymmetry. Development 127:167–175

Pellas TC, Weiss L (1990) Deep splenic lymphatic vessels in the mouse: a route of splenic exit for recirculating lymphocytes. Am J Anat 187:347–354

Qin J, Konno H, Ohshima D, Yanai H, Motegi H, Shimo Y, Hirota F, Matsumoto M, Takaki S, Inoue J, Akiyama T (2007) Developmental stage-dependent collaboration between the TNF

receptor-associated factor 6 and lymphotoxin pathways for B cell follicle organization in secondary lymphoid organs. J Immunol 179:6799–6807
Robert J, Ohta Y (2009) Comparative and developmental study of the immune system in Xenopus. Dev Dyn 238:1249–1270
Roberts CW, Shutter JR, Korsmeyer SJ (1994). Hox11 controls the genesis of the spleen. Nature 368:747–749
Satoh T, Sakurai E, Tada H, Masuda T (2009) Ontogeny of reticular framework of white pulp and marginal zone in human spleen: immunohistochemical studies of fetal spleens from the 17th to 40th week of gestation. Cell Tissue Res 336:287–297
Schneider A, Mijalski T, Schlange T, Dai W, Overbeek P, Arnold HH, Brand T (1999) The homeobox gene NKX3.2 is a target of left-right signalling and is expressed on opposite sides in chick and mouse embryos. Curr Biol 9:911–914
Sock E, Rettig SD, Enderich J, Bösl MR, Tamm ER, Wegner M (2004) Gene targeting reveals a widespread role for the high-mobility-group transcription factor Sox11 in tissue remodeling. Mol Cell Biol 24:6635–6644
Steiniger B, Barth P, Herbst B, Hartnell A, Crocker PR (1997) The species-specific structure of microanatomical compartments in the human spleen: strongly sialoadhesin-positive macrophages occur in the perifollicular zone, but not in the marginal zone. Immunology 92:307–316
Steiniger B, Barth P (2000) Microanatomy and function of the spleen. Adv Anat Embryol Cell Biol 151:III–IX, 1–101
Steiniger B, Barth P, Hellinger A (2001) The perifollicular and marginal zones of the human splenic white pulp: do fibroblasts guide lymphocyte immigration? Am J Pathol 59:501–512
Steiniger B, Ulfig N, Risse M, Barth PJ (2007a) Fetal and early post-natal development of the human spleen: from primordial arterial B-cell lobules to a non-segmented organ. Histochem Cell Biol 128:205–215
Steiniger B, Stachniss V, Schwarzbach H, Barth PJ, Steiniger B, Stachniss V, Schwarzbach H, Barth PJ (2007b) Phenotypic differences between red pulp capillary and sinusoidal endothelia help localizing the open splenic circulation in humans. Histochem Cell Biol 128:391–398
Takimoto T, Takahashi K, Sato K, Akiba Y (2005) Molecular cloning and functional characterizations of chicken TL1A. Dev Comp Immunol 29:895–905
Tierens A, Delabie J, Michiels L, Vanderberghe P, De Wolf-Petters C (1999) Marginal-zone B cells in the human lymph node and spleen show somatic hypermutations and display clonal expansion. Blood 93:226–234
Tischendorf F (1985) On the evolution of the spleen. Experientia 41:145–152
Vellguth S, von Gaudecker B, Müller-Hermelink HK (1985) The development of the human spleen. Ultrastructural studies in fetuses from the 14th to 24th week of gestation. Cell Tissue Res 242:579–592
Vigliano FA, Bermúdez R, Quiroga MI, Nieto JM (2006) Evidence for melano-macrophage centres of teleost as evolutionary precursors of germinal centres of higher vertebrates: an immunohistochemical study. Fish Shellfish Immunol 21:467–471
Vondenhoff MF, Desanti GE, Cupedo T, Bertrand JY, Cumano A, Kraal G, Mebius RE, Golub R (2008) Separation of splenic red and white pulp occurs before birth in a LTalphabeta-independent manner. J Leukoc Biol 84:152–161
Wang CC, Biben C, Robb L, Nassir F, Barnett L, Davidson NO, Koentgen F, Tarlinton D, Harvey RP (2000) Homeodomain factor Nkx2–3 controls regional expression of leukocyte homing coreceptor MAdCAM-1 in specialized endothelial cells of the viscera. Dev Biol 224:152–167
Wang J, Adelson DL, Yilmaz A, Sze SH, Jin Y, Zhu JJ (2005) Genomic organization, annotation, and ligand-receptor inferences of chicken chemokines and chemokine receptor genes based on comparative genomics. BMC Genomics 6:45–62
Wen L, Shinton SA, Hardy RR, Hayakawa K (2005) Association of B-1 B cells with follicular dendritic cells in spleen. J Immunol 174:6918–6926

Withers DR, Kim MY, Bekiaris V, Rossi SW, Jenkinson WE, Gaspal F, McConnell F, Caamano JH, Anderson G, Lane PJ (2007) The role of lymphoid tissue inducer cells in splenic white pulp development. Eur J Immunol 37:3240–3245

Yasuda M, Taura Y, Yokomizo Y, Ekino S. (1998) A comparative study of germinal center: fowls and mammals. Comp Immunol Microbiol Infect Dis 21:179–189

Zapata A, Diez B, Cejalvo T, Gutiérrez-de Frías C, Cortés A (2006) Ontogeny of the immune system of fish. Fish Shellfish Immunol 20:126–136

Zhao Y, Pan-Hammarström Q, Zhao Z, Hammarström L (2005) Identification of the activation-inducedcytidine deaminase gene from zebrafish: an evolutionary analysis. Dev Comp Immunol 29:61–71

Zhao Y, Pan-Hammarström Q, Yu S, Wertz N, Zhang X, Li N, Butler JE, Hammarström L (2006) Identification of IgF, a hinge-region-containing Ig class, and IgD in Xenopus tropicalis. Proc Natl Acad Sci USA 103:12087–12092

Zindl CL, Kim TH, Zeng M, Archambault AS, Grayson MH, Choi K, Schreiber RD, Chaplin DD (2009) The lymphotoxin LTalpha(1)beta(2) controls postnatal and adult spleen marginal sinus vascular structure and function. Immunity 30:408–420

# Chapter 12
# Formation and Function of White Pulp Lymphocyte Rich Areas of Spleen

**Peter J.L. Lane, Fiona M McConnell, and David Withers**

**Abstract** The spleen is two organs in one: the red pulp where fixed tissue macrophages remove effete red blood cells and pathogens, and the white pulp areas within which adaptive lymphocyte-dependent immune responses evolve and are maintained. In this section, we review the signals and cells that regulate the development of the white pulp areas, and how this enables the development and maintenance of immune responses. We particularly emphasize the role of lymphoid tissue inducer cells in establishing and maintaining immune responses in the spleen.

## 12.1 Anatomy of the Splenic White Pulp Areas

The splenic vasculature is supported by a network of fibers connected to the splenic capsule. The white pulp lymphocyte rich areas of the spleen ensheath the small arteriolar branches of the splenic artery that terminate in a low pressure sinusoidal system, the marginal sinus (MS), or connect directly to the red pulp (Fig. 12.1). Although the anatomy varies slightly between rodents and primates (Mebius and Kraal 2005), the MS is surrounded by a discrete lymphocyte rich area, the MZ, which is populated by a distinct subset of MZ B cells. Normally, blood that enters in the MS filters through the MZ into the red pulp macrophage lined sinusoids of the spleen.

## 12.2 The Splenic Conduit and the Fibroreticular Cells That Form It

The organization of splenic B and T cells into discrete B follicles and T cell areas depends on the splenic conduit: this is a network of fibers connecting the MS with the central arterioles that is ensheathed by fixed fibroreticular stromal

P.J.L. Lane (✉), F.M. McConnell, and D. Withers
MRC Centre for Immune Regulation, Institute for Biomedical Research, Birmingham Medical School, Birmingham, UK
e-mail: p.j.l.lane@bham.ac.uk

P. Balogh (ed.), *Developmental Biology of Peripheral Lymphoid Organs*,
DOI 10.1007/978-3-642-14429-5_12, 

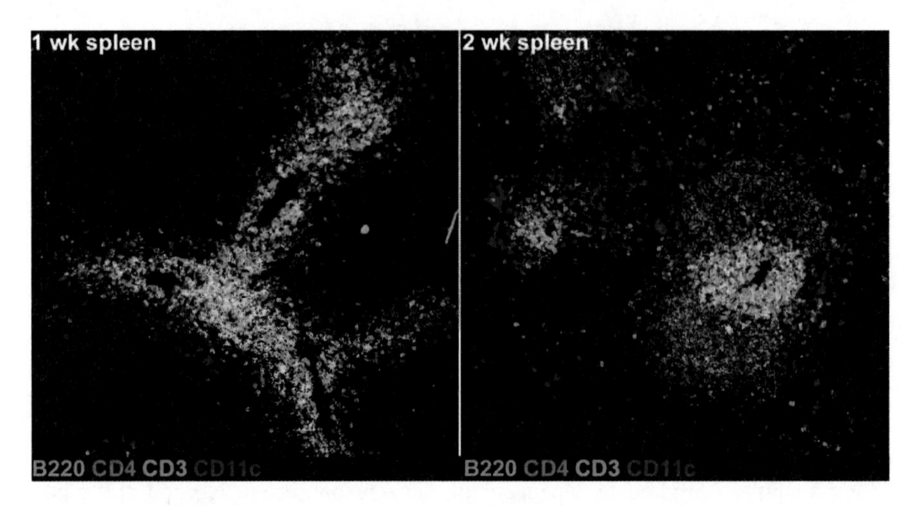


**Fig. 12.1** Hemopoetic reorganization in postnatal spleen in mice. In 1-week old spleen (left), the B cells (B220+) and T cells (CD3+) intermingle, with a slight dominance of B cells at the peripheral part of developing white pulp, where small clusters of CD4+/CD3− cells and CD11c+ cells can be observed. The central region of immature white pulp is composed mainly of T cells, and to a smaller degree, loosely organized dendritic cells (DCs) expressing CD11c. In 2-week old spleen (right) B cell rich follicles and T cell PALS clearly segregate, with a substantial accumulation of CD11c+ DCs in the latter compartment

cells (FRC) (Nolte et al. 2003). The conduit system is present in both B follicles and the T zone. In murine spleen, the FRCs of the B follicle are phenotypically different (express the marker BP3) (Ngo et al. 1999) from those in the T zone, which express the transmembrane mucin, podoplanin (Farr et al. 1992) (Fig. 12.2). Both the B and T zone conduits are permeable to small molecules, notably chemokines. The FRCs in B follicles secrete the B cell attracting chemokine, CXCL13, into the B zone conduit, whereas T zone FRCs secrete the T zone chemokines, CCL19 and CCL21. Chemokines secreted at the abluminal FRC surface effectively distribute chemokines throughout the B and T conduit network (Nolte et al. 2003), probably enabling even distribution. It is not certain that the B and T cell conduits are connected, but under normal circumstances, CXCL13 is restricted to the B follicle, and CCL19 and CCL21 to the T zone.

Intravenous injection of low molecular weight proteins and polysaccharides fills the splenic conduit system in a few seconds from both the MS and the central arteriole (Nolte et al. 2003). This indicates direct communications between the MS and central arterioles and the conduit. Although the conduit can transport low molecular weight immunogens that can subsequently be taken up and presented by dendritic cells in the T zone, it is not clear whether this is a fundamental function of the conduit. What is clear is that chemokines secreted by both B and T zone FRCs can be transported inside the conduits.

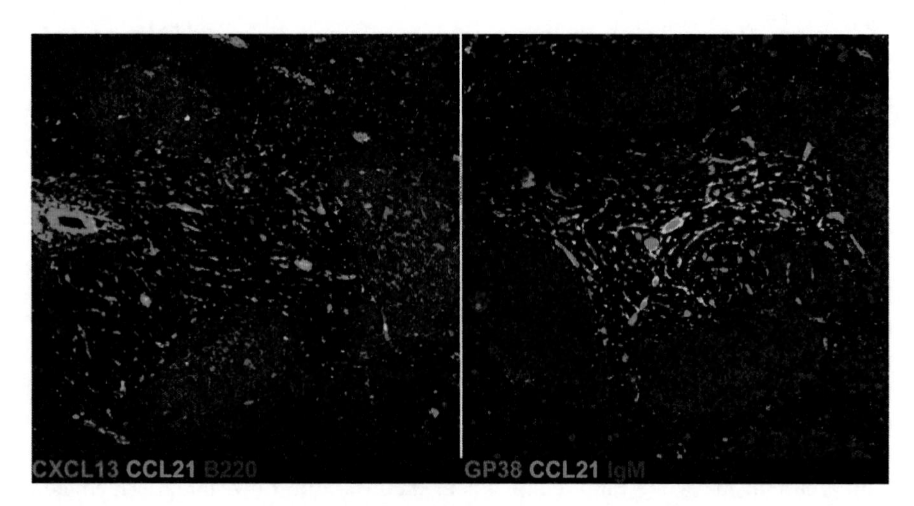


**Fig. 12.2** Reticular fibroblastic and homeostatic chemokine architecture of adult spleen. B cells ($B220^+$) accumulate around follicular stromal cells that produce CXCL13 chemokine (*right*), whereas the PALS-associated reticular fibroblastic cells produce CCL21 (*left*). These latter cells can be identified by their podoplanin expression/$gp38^+$ expression (*right*).

## 12.3 Homeostatic Chemokines Organize the White Pulp Areas

The homeostatic B zone chemokine (CXCL13), distributed in the B cell conduit, and the T zone chemokines (CCL19, CCL21) (Cyster 2003), distributed in the T zone conduit, are required for the formation of B follicles and the T zone. Although required for the formation of B follicles, B/T segregation is not dependent on expression of CXCL13, as B and T cells segregate normally in mice deficient in CXCR5, the receptor for CXCL13 (Forster et al. 1996).

Of the two CCR7 ligands, murine CCL21 is produced in greater abundance (100-fold greater protein levels than CCL19) (Luther et al. 2002). Furthermore, ectopic expression of CCL21 induces larger and more organized infiltrates than CCL19, so it is functionally and quantitatively more important than CCL19.

The main structural difference between CCL19 and CCL21 is that the latter has an extra 32 amino acid C-terminus containing basic aminoacids (Hirose et al. 2002; Ueno et al. 2002). Through this basic motif it binds negatively charged structures, including the highly glycosylated mucin-type protein podoplanin (Kerjaschki et al. 2004) expressed on T zone stroma (Farr et al. 1992). As a consequence of its immobilization, CCL21 is less chemotactic than soluble CCL19 (Hirose et al. 2002; Ueno et al. 2002). The result of this immobilization is that unlike CCL19, CCL21 does not desensitize its receptor (Forster et al. 2008). Furthermore, immobilization of CCL21 in vitro (simulating immobilization on podoplanin-expressing T zone stroma) stimulates integrin-independent T cell chemokinesis (Lammermann et al. 2008; Woolf et al. 2007), a process fundamental to the promotion of interactions between T cells migrating on T zone stromal cells (Bajenoff et al. 2006) and DCs in the T zone.

A similar situation exists for CXCL13 in the B follicle (Allen et al. 2007). The chemokinesis of follicular B cells is stimulated by engagement of CXCR5 by CXCL13 presented on B follicular stromal cells.

## 12.4 B and T Cell Recirculation Through the Spleen

Both B and T lymphocytes recirculate between blood and lymph (Gowans 1959). In the spleen, B and T cells enter in the MS. Both CCR7 and CXCR5 ligands are presented on the endothelium of splenic blood vessels (Nolte et al. 2003). Under flow conditions, chemokine signals stimulate adhesion (Alon and Ley 2008), so it is possible that signaling through homeostatic chemokines in splenic blood vessels stimulates adhesion and then entry of both B and T cells from the MS into the white pulp areas. In the white pulp areas, where there is little flow, integrin-mediated adhesion is switched off (Woolf et al. 2007). Studies show that lymphocytes track in association with the conduit network of FRCs that present homeostatic chemokines, and this enables rare antigen-specific T cells to find antigen-presenting dendritic cells (DCs).

In the absence of antigen-mediated triggering, lymphocytes leave the splenic white pulp by upregulating their expression of sphingosine-1-phosphate (S1P) receptor 1 (SIP1) (Matloubian et al. 2004), the ligand for which is present in blood, and splenic lymphocytes probably leave the white pulp areas, to migrate through the MZ bridging channels into the splenic red pulp, hence returning to the blood. However, in antigen-triggered cells, CD69 is upregulated and antagonizes responses to SIP1 signaling, leading to the retention of activated lymphocytes (Shiow et al. 2006).

## 12.5 Anatomy of T Cell-Dependent Antibody Responses

The segregation of B and T cells is functionally related to the capacity of mammals to make high affinity antibodies. During immune responses to T cell-dependent (TD) protein antigens, CD4 T cells that encounter antigen presented by T zone dendritic cells upregulate expression of CXCR5 and migrate to the outer T zone and interface between the B follicle and the T zone. Similarly, B cells triggered by native antigen through their B cell receptor upregulate the T zone chemokine CCR7 (Cyster et al. 1994; Reif et al. 2002). The coexpression of CXCR5 and CCR7 on activated B and T cells localizes both cell types at the B/T interface, facilitating their interaction, and thus the provision of T cell help for B cells. Following cognate interactions between primed B and T cells, two distinct B cell processes evolve.

One consequence of T/B collaboration is the generation of rapidly proliferating plasmablasts that differentiate into short-lived plasma cells; these migrate to the red pulp of the spleen and secrete low affinity nonclass switched and class switched antibodies (Liu et al. 1991). However, the second process is fundamentally much more important. A distinct subset of CD4 follicular T helper cells ($T_{FH}$) down-regulates CCR7 and migrates into B follicles, where it drives the proliferation of follicular B cells to form germinal centers (GCs) (MacLennan 1994), in which the same $T_{FH}$ execute the iterative selection of somatically mutating B cells. The products of this process are: long-lived plasma cells bearing high affinity class switched immunoglobulin receptors, which migrate to the bone marrow (Ho et al. 1986; Slifka et al. 1998); and memory B cells, which recirculate and also colonize the MZ (Liu et al. 1988).

## 12.6 The Splenic Marginal Zone

Unlike follicular B cells, MZ B cells do not recirculate but are derived from recirculating cells (Kumararatne et al. 1981). Their retention at the interface between the red and white pulp is integrin- (Lu and Cyster 2002) and SIP1-dependent (Cinamon et al. 2004). The latter study suggested that SIP1 signals from blood counterbalanced those through CXCR5 signals from CXCL13 expressing FRCs, localizing MZ B cells at the interface between the red and white pulp areas.

## 12.7 MZ B Cell Transport of Antigen onto FDCs in B Follicles

MZ B cells, by virtue of their location, filter the blood that enters the MS. One functional of MZ B cells that has been clearly identified is the transport of antigens that have fixed complement onto FDCs (Gray et al. 1984), enabling the initiation of FDC-associated B cell activation in B follicles. Activation of MZ B cells by immune complexes upregulates the expression of CD69 and downregulates the expression of SIP1, attenuating their responsiveness to SIP1 signals in the blood and enhancing their response to CXCL13 in B follicles, thus enabling their migration into B follicles (Rubtsov et al. 2008). Elegant studies have indicated that this shuttling of MZ B cells between MZ and B follicles is a continuous process resulting in the efficient transport of antigens to FDCs through the binding of immune complexes (Cinamon et al. 2008).

## 12.8 MZ B Cells and Antibody Responses

By virtue of their perfusion with the blood that enters the MZ, MZ B cells are also ideally situated to be activated through their antigen-specific receptors by blood-borne pathogens. Indeed, early studies implicated MZ B cells in responses to T cell-independent (TI) antigens, for example, those expressed by encapsulated bacteria

(Lane and MacLennan 1986) and more elegant studies have confirmed this (Martin et al. 2001). MZ B cells are preactivated, and following antigen encounter differentiate rapidly into antibody producing cells (Oliver et al. 1997; Oliver et al. 1999).

However, MZ B cells are not only linked to TI immune responses. As stated above, memory B cells colonize the MZ following immunization with TD protein antigens as well (Liu et al. 1988) and following activation by antigen during memory immune responses, they proceed rapidly to the secretion of memory high affinity antibodies. This does not necessarily require T cell help (Hebeis et al. 2004).

## 12.9 Splenic Phylogeny Reveals That White Pulp Organization Evolved in the Context of the Capacity to Make and Sustain High Affinity Antibody Responses

Although all jawed vertebrates have a spleen, there are fundamental differences in the anatomy of the red and white pulp areas (Zapata and Ameimiya 2000). In bony fish, there is little evidence of red pulp/white pulp segregation, nor are B and T cells separated into distinct areas. In higher vertebrates, like amphibians and reptiles, there is evidence of red pulp/white pulp segregation, as well as of increasing organization of white pulp areas, although these animals are unable to form B cell GC after immunization. GCs are only seen in endothermic birds and mammals, and only the latter have a properly formed MZ.

The development of organization in the spleen maps to the capacity to make and sustain GC-dependent high affinity class-switched memory antibody responses, which are strikingly absent in lower vertebrates (Zapata and Ameimiya 2000), and we have suggested that these functions coevolved (Lane et al. 2005).

## 12.10 Recently Evolved Tumor Necrosis Family Member Ligands Control the Development of Organization and CD4 Memory in the Spleen

The TNF family of receptor and ligands plays a pivotal role in the development and organization of the immune system. For example, the TNF family member, CD4 T cell CD40-ligand (TNFSF5) through its interaction with B cell CD40 (TNFRSF5) plays the pivotal role in T cell help for B cells (Banchereau et al. 1994; Grewal and Flavell 1996), and this function is conserved in all vertebrates (Gong et al. 2009). In addition, other TNF-ligands (BAFF and APRIL) implicated in B (Mackay and Browning 2002) and plasma cell (Belnoue et al. 2008; Benson et al. 2008) survival, respectively, are found in all vertebrates (Glenney and Wiens 2007).

However, there are several TNF family members only present in higher vertebrates. Gene duplication is the principle mechanism that facilitates the evolution of new genes (Levy et al. 2007). Iterated over millions of generations as species diverge, diversification of duplicated genes provides the genetic fuel for the acquisition of new genes that control new functions; four genes responsible for lymphoid tissue organization and CD4 memory are amongst those TNF family members present only in higher vertebrates (Glenney and Wiens 2007). These are the lymphotoxin genes (LTα and LTβ), OX40-ligand (OX40L) (TNFSF4), and CD30L (TNFSF8).

The receptors for these genes are located in two TNF clusters at the tip of human chromosome 1 (OX40 and CD30) and 12 (LTβR and TNFR1) (Table 12.1). In addition to TNFR family members, the two gene clusters each contain a closely related serine protease, MASP2 (Chromosome 1) and C1rs (Chromosome 12), which catalyze the activation of complement component 4 via the mannan binding and the classical pathways, respectively. It seems likely that the 2 TNF clusters may have arisen from an ancestral chromosomal duplication that predated the evolution of jawed vertebrates, as originally proposed by Ohno (Ohno 1970), and for which there is direct evidence (Putnam et al. 2008). In support of this, although teleosts lack some TNF members, both gene clusters are present in fish genomes (http://www.ensembl.org).

## 12.11 Splenic Lymphoid Tissue Inducer Cells Express High Levels of the Newly Evolved TNF Ligands, LTα and LTβ, CD30L, and OX40L

LTi were first described in detail in developing embryonic lymph nodes (Mebius et al. 1997). They are characterized by their expression of many different TNF ligands. However, they do not express any of the TNF members linked with B cell activation and survival (CD40, BAFF, and APRIL) (Kim et al. 2006). The expressed

**Table 12.1** TNF receptor clusters linked with lymphoid tissue organization and CD4 memory

| TNF cluster 1 | Chromosome 1 position (MB) | TNF cluster 2 | Chromosome 12 position (MB) |
|---|---|---|---|
| GITR | 1.13 | TNFR1* | 12:6.3 |
| OX40** | 1.14 | LTβR** | 12:6.36 |
| HVEM*** | 2.5 | CD27 | 12:6.4 |
| DR3* | 6.4 | CD4 | 12:6.7 |
| 4–1BB | 7.9 | C1rs$ | 12:7.1 |
| MASP2$ | 11 | AID | 12:8.6 |
| CD30 | 12 | | |
| TNFR2*** | 12.1 | | |

Human TNF receptors on Chromosome 1 and 12. Gene marked in black letters show TNF receptors present in all vertebrates. Genes marked in bold show TNF receptors whose ligands are only present in higher vertebrates. Genes marked in with $ sign show the 2 serine proteases that catalytically cleave complement component,C4. Asterisked TNF receptors are close paralogues. Data from http://www.ensembl.org

molecules include TNFL shared with primitive vertebrates [TNFα(TNFSF2), LIGHT (TNFSF14), and TRANCE (TNFSF11A)] in addition to the more recently evolved TNFL; adult and embryonic LTi both express high levels of the lymphotoxins, LTα (TNFSF1) and LTβ (TNFSF3), and both can be induced to express high levels of OX40L (TNFSF4) and CD30L (TNFSF8) (Kim et al. 2006).

LTi development depends on the splice variant of the transcription factor, retinoic acid orphan receptor gamma (RORγt) (Eberl and Littman 2004; Eberl et al. 2004; Sun et al. 2000). Their role in the lymphotoxin-dependent induction of both conventional lymph nodes and Peyer's patches (Eberl and Littman 2003; Mebius 2003) but also inducible lymphoid tissues (Eberl 2005) is well established, but the spleen looks normally organized in RORγt-deficient mice, so splenic development and organization does not depend on conventional LTi.

Although conventional $ROR\gamma t^+$ LTi are not essential for the formation of white pulp areas of the spleen (Kurebayashi et al. 2000; Sun et al. 2000), conventional LTi are present in both the fetal (Vondenhoff et al. 2008; Withers et al. 2007) and adult (Kim et al. 2003; Scandella et al. 2008) spleen where they attach to fixed FRC in both B and T cell areas (Kim et al. 2007). The expression of lymphotoxins by splenic LTi cannot induce expression of CXCL13 and B follicle formation, which is dependent on B cell lymphotoxins (Fu et al. 1998; Gonzalez et al. 1998), but they can induce expression of T zone chemokines in the absence of lymphotoxin expression by lymphocytes. Furthermore, although originally only thought to be functional in the embryo, adult splenic LTi have been implicated in the repair of spleen following viral damage (Scandella et al. 2008), a function analogous to the one they perform in fetal life.

Both murine fetal and adult splenic LTi are heterogeneous with regard to their expression of homeostatic chemokine receptors, CXCR5 and CCR7 (Kim et al. 2008), consistent with the three locations where they are found in adult murine spleen: associated with B FRCs ($CXCR5^+CCR7^-$), at the boundary between the B and T cell areas ($CXCR5^+CCR7^+$) (Kim et al. 2003), and in the central T zone areas adjacent to T zone FRCs ($CXCR5^-CCR7^+$) (Kim et al. 2007).

Interactions between LTi and the underlying stroma, as well as with DCs and T cells, can readily be identified in mouse splenic T zones (Kim et al. 2007), as can associations between primed CD4 T cells and LTi in B follicles and at the B/T (Kim et al. 2003). These associations are consistent with provision by LTi of signals to CD4 T cells at the points of initiation of CD4 responses and of subsequent help for B cells.

## 12.12 High Affinity Class Switched Antibodies Depend on LTα- and LTβ-induced Organization

LTα and LTβ by binding to the TNF receptor (TNFRSF1A) and the lymphotoxin beta receptor (LTβR) provide the key signals for the development of organized secondary lymphoid tissues (Fu and Chaplin 1999), although signals through CD30 play additional roles in lymphoid tissue organization (Bekiaris et al. 2009; Bekiaris et al. 2007). B cell LTβR signals are particularly important for induction of CXCL13 and B follicle formation, whereas in the absence of lymphotoxin signals, a residual function for CD30 signals in the induction of the T zone chemokine, CCL21, can be demonstrated (Bekiaris et al. 2009).

The formation of discrete follicles enables CD4 T cells to orchestrate the CD40-dependent proliferation, somatic mutation, and selection of B cells within GCs, with the consequent production of high affinity class switched antibody responses. Consequently, lymphotoxin-deficient mice, which lack B follicles, produce neither high affinity IgG in the systemic circulation (Fu et al. 1997) nor IgA at mucosal surfaces (Kang et al. 2002). In contrast, T-dependent primary low affinity nonswitched IgM responses do not depend on organized lymphoid structures and B follicles.

Antibody responses in disorganized (lymphotoxin deficient) mice mimic the antibody immune system found in lower vertebrates, which lack these genes, despite having the activation-induced cytidine deaminase (AID) (Rogozin et al. 2007) that is required for both class switching and somatic mutation (Muramatsu et al. 2000), intact T cell help through CD40 (Gong et al. 2009), and somatic mutation of immunoglobulin genes (Wilson et al. 1992). This demonstrates the key importance of the lymphotoxin-dependent B follicle as the structure for generation of high affinity antibodies, consistent with the idea that the driving force for the selection of organized lymphoid structures was the acquisition of a mechanism to select efficiently the B cell precursors of high affinity class switched antibodies.

The selective advantage conferred by high affinity antibodies, in combination with the investment involved in their production, must have applied potent selective pressure for mechanisms to maintain them, now evolved as B and T cell memory. Under normal circumstances, memory B and T cells collaborate with each other in the outer T zone of secondary lymphoid structures (Liu et al. 1991). Organization is required for this interaction, as memory B and T cells fail to find each other in the absence of organization (Fu et al. 2000). This implies that lymphoid tissue organization was a prerequisite for the development of memory antibody responses.

## 12.13 Adult LTi Express High Levels of the TNF-ligands, OX40L and CD30L, and These Molecules Are Linked with the Development of CD4 Memory

In addition to the lymphotoxin genes required for organization, LTi can express high levels of OX40L and CD30L (Lane et al. 2005). These TNF ligands are not exclusive to LTi, but LTi isolated from adult T cell-deficient mice express high levels of these molecules (Kim et al. 2003). This contrasts with the situation in embryonic LTi, where expression of these proteins is absent (Kim et al. 2005), but is upregulated upon transfer into an adult environment (Kim et al. 2006).

We have accumulated evidence suggesting that activated DCs producing proinflammatory cytokines upregulate expression of OX40L, for example, the soluble

TNF ligand, TL1A (Meylan et al. 2008), strongly upregulates expression of OX40L on embryonic and adult LTi in vitro (Kim et al. 2006).

The receptors for OX40 and CD30 are primarily expressed on antigen-activated T cells (Croft 2003; Watts 2005). Because of our evidence that primed CD4 T cells could interact directly with LTi (Kim et al. 2003), which could express high levels of the ligands, we made mice singly deficient in OX40 and CD30, or deficient in both molecules. Individual contributions of OX40 and CD30 signals to CD4 T cell memory were identified (Gaspal et al. 2005; Kim et al. 2003) but the striking result was that when both signaling pathways were removed, CD4 memory generation was abrogated.

The importance of these molecules was first demonstrated for CD4 memory for antibody responses (Gaspal et al. 2005), but is also true for Th1 CD4 memory (Gaspal et al. 2008). Although the expression of OX40L and CD30L is not exclusive to LTi, we have recent evidence that LTi provision of these signals is crucial, as CD4 T cell priming and recall responses are very poor in ROR-γt-deficient mice lacking LTi (Withers and Lane, unpublished observations).

## 12.14 Ontogeny of Splenic White Pulp Areas in Relation to LTi Expression of Lymphotoxins, OX40L and CD30L

In the embryonic murine spleen, LTi are found from embryonic day (E) 13 and increase in numbers on E14 and E15 (Withers et al. 2007). By E15-E16, LTi are clustered specifically around $CD31^+$ (PECAM-1) blood vessels, where they interact with VCAM-$1^+$MAdCAM-$1^-$ stromal FRC. Although clustering of LTi with FRC was independent of lymphotoxin signals, upregulation of VCAM-1 on FRC was substantially lymphotoxin dependent. Therefore, as described for developing LNs in the embryo (Mebius 2003), LTis are amongst the first haematopoietic cells detected in the spleen and they cluster with a specific subset of stromal cells around blood vessels, the site where white pulp areas form in normal spleen.

The development of nascent white pulp areas in lymphocyte-deficient $RAG^{-/-}$ spleen shows LTi clustering with VCAM-$1^+$ stroma around the central arterioles. In the embryonic spleens of $Rag^{-/-}$ mice also deficient in the common gamma chain of the interleukin 2 receptor, which have only ~10% of the normal numbers of splenic LTi (Kim et al. 2005), there are proportionally fewer white pulp areas. In summary, LTi clustering around blood vessels in the spleen in the embryo identifies areas that are subsequently colonized by T cells and DCs in a lymphotoxin-dependent manner after birth. However, the development of splenic white pulp areas in mice lacking LTi (RORγ-deficient mice) indicates that at least in spleen, LTi are not essential.

Lymphocytes start to appear in the spleen after birth but are poorly segregated initially. The first chemokine to appear is CCL21 and this is coincident with upregulation of podoplanin on FRCs and the development of a ring structure with a core of T cells surrounded by B cells but no discrete CXCL13 expressing B follicles (Withers et al. 2007). These ring structures are similar to those observed in mice deficient in CXCR5 (CXCL13 signals). Evidence that CCL21 is immobilized by binding to podoplanin (Kerjaschki et al. 2004) could explain this initial B/T segregation. Both lymphotoxin (Ngo et al. 2001) and CD30 (Bekiaris et al. 2007) signals, whose ligands are both expressed by LTi, are required for podoplanin expression; therefore, LTi, by virtue of their association with FRCs, could provide these signals. In the embryo and neonatal period, LTi do not express either CD30L or OX40L (Kim et al. 2005), but the upregulation of podoplanin is temporarily associated with LTi CD30L expression (Withers et al. 2007).

The development of splenic B follicles occurs a few days later and is dependent on B cell lymphotoxin expression. The final B cell structure to emerge is the MZ, which starts to develop from 1 week of age and is also dependent on lymphotoxin signals (Nolte et al. 2004).

## 12.15 Summary

The organized white pulp structures of mammalian spleen evolved from poorly segregated lymphocyte rich structures in primitive vertebrates. The development of organization is linked with functional lymphotoxin genes and particularly the LTβR. The evolution of a distinct and segregated B follicle is linked in phylogeny with the capacity to make GCs, in which B cell mutants generating high affinity antibodies are efficiently selected by follicular CD4 T cells, and to sustain them through the development of memory T and B cells. Splenic LTi though their expression of the newly evolved TNF-ligands, OX40L and CD30L, are linked with the development of both high affinity antibodies and memory.

**Acknowledgments** This work was supported by a Wellcome Trust Programme Grant to PL.

## References

Allen CD, Okada T, Tang HL, Cyster JG (2007) Imaging of germinal center selection events during affinity maturation. Science 315:528–531

Alon R, Ley K (2008) Cells on the run: shear-regulated integrin activation in leukocyte rolling and arrest on endothelial cells. Curr Opin Cell Biol 20:525–532

Bajenoff M, Egen JG, Koo LY, Laugier JP, Brau F, Glaichenhaus N, Germain RN (2006) Stromal cell networks regulate lymphocyte entry, migration, and territoriality in lymph nodes. Immunity 25:989–1001

Banchereau J, Bazan F, Blanchard D, Briere F, Galizzi JP, van Kooten C, Liu YJ, Rousset F, Saeland S (1994) The CD40 antigen and its ligand. Annu Rev Immunol 12:881–922

Bekiaris V, Gaspal F, Kim MY, Withers DR, McConnell FM, Anderson G, Lane PJ (2009) CD30 is required for CCL21 expression and CD4 T cell recruitment in the absence of lymphotoxin signals. J Immunol 182:4771–4775

Bekiaris V, Withers D, Glanville SH, McConnell FM, Parnell SM, Kim MY, Gaspal FM, Jenkinson E, Sweet C, Anderson G, Lane PJ (2007) Role of CD30 in B/T segregation in the spleen. J Immunol 179:7535–7543

Belnoue E, Pihlgren M, McGaha TL, Tougne C, Rochat AF, Bossen C, Schneider P, Huard B, Lambert PH, Siegrist CA (2008) APRIL is critical for plasmablast survival in the bone marrow and poorly expressed by early-life bone marrow stromal cells. Blood 111:2755–2764

Benson MJ, Dillon SR, Castigli E, Geha RS, Xu S, Lam KP, Noelle RJ (2008) Cutting edge: the dependence of plasma cells and independence of memory B cells on BAFF and APRIL. J Immunol 180:3655–3659

Cinamon G, Matloubian M, Lesneski MJ, Xu Y, Low C, Lu T, Proia RL, Cyster JG (2004) Sphingosine 1-phosphate receptor 1 promotes B cell localization in the splenic marginal zone. Nat Immunol 5:713–720

Cinamon G, Zachariah MA, Lam OM, Foss FW, Jr., Cyster JG (2008) Follicular shuttling of marginal zone B cells facilitates antigen transport. Nat Immunol 9:54–62

Croft M (2003) Co-stimulatory members of the TNFR family: keys to effective T-cell immunity? Nat Rev Immunol 3:609–620

Cyster JG (2003) Lymphoid organ development and cell migration. Immunol Rev 195:5–14

Cyster JG, Hartley SB, Goodnow CC (1994) Competition for follicular niches excludes self-reactive cells from the recirculating B-cell repertoire. Nature 371:389–395

Eberl G (2005) Opinion: Inducible lymphoid tissues in the adult gut: recapitulation of a fetal developmental pathway? Nat Rev Immunol 5:413–420

Eberl G, Littman DR (2003) The role of the nuclear hormone receptor RORgammat in the development of lymph nodes and Peyer's patches. Immunol Rev 195:81–90

Eberl G, Littman DR (2004) Thymic origin of intestinal alphabeta T cells revealed by fate mapping of RORgammat+ cells. Science 305:248–251

Eberl G, Marmon S, Sunshine MJ, Rennert PD, Choi Y, Littman DR (2004) An essential function for the nuclear receptor RORgamma(t) in the generation of fetal lymphoid tissue inducer cells. Nat Immunol 5:64–73

Farr AG, Berry ML, Kim A, Nelson AJ, Welch MP, Aruffo A (1992) Characterization and cloning of a novel glycoprotein expressed by stromal cells in T-dependent areas of peripheral lymphoid tissues. J Exp Med 176:1477–1482

Forster R, Davalos-Misslitz AC, Rot A (2008) CCR7 and its ligands: balancing immunity and tolerance. Nat Rev Immunol 8:362–371

Forster R, Mattis AE, Kremmer E, Wolf E, Brem G, Lipp M (1996) A putative chemokine receptor, blr1, directs b-cell migration to defined lymphoid organs and specific anatomic compartments of the spleen. Cell 87:1037–1047

Fu YX, Chaplin DD (1999) Development and maturation of secondary lymphoid tissues. Annu Rev Immunol 17:399–433

Fu YX, Huang G, Wang Y, Chaplin DD (1998) B lymphocytes induce the formation of follicular dendritic cell clusters in a lymphotoxin alpha-dependent fashion. J Exp Med 187:1009–1018

Fu YX, Huang G, Wang Y, Chaplin DD (2000) Lymphotoxin-alpha-dependent spleen microenvironment supports the generation of memory B cells and is required for their subsequent antigen-induced activation. J Immunol 164:2508–2514

Fu YX, Molina H, Matsumoto M, Huang G, Min J, Chaplin DD (1997) Lymphotoxin-alpha (LTalpha) supports development of splenic follicular structure that is required for IgG responses. J Exp Med 185:2111–2120

Gaspal F, Bekiaris V, Kim M, Withers D, Bobat S, MacLennan I, Anderson G, Lane PJ, Cunningham AF (2008) Critical synergy of CD30 and OX40 signals in CD4 T cell homeostasis and Th1 immunity to salmonella. J Immunol 180:2824–2829

Gaspal FM, Kim MY, McConnell FM, Raykundalia C, Bekiaris V, Lane PJ (2005) Mice deficient in OX40 and CD30 signals lack memory antibody responses because of deficient CD4 T cell memory. J Immunol 174:3891–3896

Glenney GW, Wiens GD (2007) Early diversification of the TNF superfamily in teleosts: genomic characterization and expression analysis. J Immunol 178:7955–7973

Gong YF, Xiang LX, Shao JZ (2009) CD154-CD40 interactions are essential for thymus-dependent antibody production in zebrafish: insights into the origin of costimulatory pathway in helper T cell-regulated adaptive immunity in early vertebrates. J Immunol 182:7749–7762

Gonzalez M, Mackay F, Browning JL, Kosco-Vilbois MH, Noelle RJ (1998) The sequential role of lymphotoxin and B cells in the development of splenic follicles. J Exp Med 187:997–1007

Gowans JL (1959) The recirculation of lymphocytes from blood to lymph in the rat. J Physiol 146:54–69

Gray D, Kumararatne DS, Lortan J, Khan M, MacLennan IC (1984) Relation of intra-splenic migration of marginal zone B cells to antigen localization on follicular dendritic cells. Immunology 52:659–669

Grewal IS, Flavell RA (1996) A central role of cd40 ligand in the regulation of cd4(+) t-cell responses. Immunol Today 17:410–414

Hebeis BJ, Klenovsek K, Rohwer P, Ritter U, Schneider A, Mach M, Winkler TH (2004) Activation of virus-specific memory B cells in the absence of T cell help. J Exp Med 199:593–602

Hirose J, Kawashima H, Swope Willis M, Springer TA, Hasegawa H, Yoshie O, Miyasaka M (2002) Chondroitin sulfate B exerts its inhibitory effect on secondary lymphoid tissue chemokine (SLC) by binding to the C-terminus of SLC. Biochim Biophys Acta 1571:219–224

Ho F, Lortan JE, MacLennan IC, Khan M (1986) Distinct short-lived and long-lived antibody-producing cell populations. Eur J Immunol 16:1297–1301

Kang HS, Chin RK, Wang Y, Yu P, Wang J, Newell KA, Fu YX (2002) Signaling via LTbetaR on the lamina propria stromal cells of the gut is required for IgA production. Nat Immunol 3:576–582

Kerjaschki D, Regele HM, Moosberger I, Nagy-Bojarski K, Watschinger B, Soleiman A, Birner P, Krieger S, Hovorka A, Silberhumer G, Laakkonen P, Petrova T, Langer B, Raab I (2004) Lymphatic neoangiogenesis in human kidney transplants is associated with immunologically active lymphocytic infiltrates. J Am Soc Nephrol 15:603–612

Kim M-Y, Anderson G, Martensson I-L, Erlandsson L, Arlt W, White A, Lane PJL (2005) OX40-ligand and CD30-ligand are expressed on adult but not neonatal $CD4^{+}CD3$- inducer cells: evidence that IL7 signals regulate CD30-ligand but not OX40-ligand expression. J Immunol 174:6686–6691

Kim MY, Gaspal FM, Wiggett HE, McConnell FM, Gulbranson-Judge A, Raykundalia C, Walker LS, Goodall MD, Lane PJ (2003) CD4(+)CD3(−) accessory cells costimulate primed CD4 T cells through OX40 and CD30 at sites where T cells collaborate with B cells. Immunity 18:643–654

Kim MY, McConnell FM, Gaspal FM, White A, Glanville SH, Bekiaris V, Walker LS, Caamano J, Jenkinson E, Anderson G, Lane PJ (2007) Function of $CD4^{+}CD3^{-}$ cells in relation to B- and T-zone stroma in spleen. Blood 109:1602–1610

Kim MY, Rossi S, Withers D, McConnell F, Toellner KM, Gaspal F, Jenkinson E, Anderson G, Lane PJ (2008) Heterogeneity of lymphoid tissue inducer cell populations present in embryonic and adult mouse lymphoid tissues. Immunology 124:166–174

Kim MY, Toellner KM, White A, McConnell FM, Gaspal FM, Parnell SM, Jenkinson E, Anderson G, Lane PJ (2006) Neonatal and adult $CD4^{+}CD3$- cells share similar gene expression profile, and neonatal cells up-regulate OX40 ligand in response to TL1A (TNFSF15). J Immunol 177:3074–3081

Kumararatne DS, Bazin H, MacLennan IC (1981) Marginal zones: the major B cell compartment of rat spleens. Eur J Immunol 11:858–864

Kurebayashi S, Ueda E, Sakaue M, Patel DD, Medvedev A, Zhang F, Jetten AM (2000) Retinoid-related orphan receptor gamma (RORgamma) is essential for lymphoid organogenesis and controls apoptosis during thymopoiesis. Proc Natl Acad Sci USA 97:10132–10137

Lammermann T, Bader BL, Monkley SJ, Worbs T, Wedlich-Soldner R, Hirsch K, Keller M, Forster R, Critchley DR, Fassler R, Sixt M (2008) Rapid leukocyte migration by integrin-independent flowing and squeezing. Nature 453:51–55

Lane PJL, Gaspal MC, Kim M-Y (2005) Two sides of a cellular coin: $CD4^+CD3^-$ cells orchestrate memory antibody responses and lymph node organisation. Nat Rev Immunol 5:655–660

Lane PJL, MacLennan ICM (1986) Impaired IgG2 anti-pneumococcal antibody responses in patients with recurrent infection and normal IgG2 levels but no IgA. Clin Exp Immunol 65:427–433

Levy S, Sutton G, Ng PC, Feuk L, Halpern AL, Walenz BP, Axelrod N, Huang J, Kirkness EF, Denisov G, Lin Y, MacDonald JR, Pang AW, Shago M, Stockwell TB, Tsiamouri A, Bafna V, Bansal V, Kravitz SA, Busam DA, Beeson KY, McIntosh TC, Remington KA, Abril JF, Gill J, Borman J, Rogers YH, Frazier ME, Scherer SW, Strausberg RL, Venter JC (2007) The diploid genome sequence of an individual human. PLoS Biol 5:e254

Liu YJ, Oldfield S, MacLennan IC (1988) Memory B cells in T cell-dependent antibody responses colonize the splenic marginal zones. Eur J Immunol 18:355–362

Liu YJ, Zhang J, Lane PJ, Chan EY, MacLennan IC (1991) Sites of specific B cell activation in primary and secondary responses to T cell-dependent and T cell-independent antigens. Eur J Immunol 21:2951–2962

Lu TT, Cyster JG (2002) Integrin-mediated long-term B cell retention in the splenic marginal zone. Science 297:409–412

Luther SA, Bidgol A, Hargreaves DC, Schmidt A, Xu Y, Paniyadi J, Matloubian M, Cyster JG (2002) Differing activities of homeostatic chemokines CCL19, CCL21, and CXCL12 in lymphocyte and dendritic cell recruitment and lymphoid neogenesis. J Immunol 169:424–433

Mackay F, Browning JL (2002) BAFF: a fundamental survival factor for B cells. Nat Rev Immunol 2:465–475

MacLennan ICM (1994) Germinal centers. Annu Rev Immunol 12:117–139

Martin F, Oliver AM, Kearney JF (2001) Marginal zone and B1 B cells unite in the early response against T-independent blood-borne particulate antigens. Immunity 14:617–629

Matloubian M, Lo CG, Cinamon G, Lesneski MJ, Xu Y, Brinkmann V, Allende ML, Proia RL, Cyster JG (2004) Lymphocyte egress from thymus and peripheral lymphoid organs is dependent on S1P receptor 1. Nature 427:355–360

Mebius RE (2003) Organogenesis of lymphoid tissues. Nat Rev Immunol 3:292–303

Mebius RE, Kraal G (2005) Structure and function of the spleen. Nat Rev Immunol 5:606–616

Mebius RE, Rennert P, Weissman IL (1997) Developing lymph nodes collect $CD4^+CD3^-$ $LTbeta^+$ cells that can differentiate to APC, NK cells, and follicular cells but not T or B cells. Immunity 7:493–504

Meylan F, Davidson TS, Kahle E, Kinder M, Acharya K, Jankovic D, Bundoc V, Hodges M, Shevach EM, Keane-Myers A, Wang EC, Siegel RM (2008) The TNF-family receptor DR3 is essential for diverse T cell-mediated inflammatory diseases. Immunity 29:79–89

Muramatsu M, Kinoshita K, Fagarasan S, Yamada S, Shinkai Y, Honjo T (2000) Class switch recombination and hypermutation require activation-induced cytidine deaminase (AID), a potential RNA editing enzyme [see comments]. Cell 102:553–563

Ngo VN, Cornall RJ, Cyster JG (2001) Splenic T zone development is B cell dependent. J Exp Med 194:1649–1660

Ngo VN, Korner H, Gunn MD, Schmidt KN, Riminton DS, Cooper MD, Browning JL, Sedgwick JD, Cyster JG (1999) Lymphotoxin alpha/beta and tumor necrosis factor are required for stromal cell expression of homing chemokines in B and T cell areas of the spleen. J Exp Med 189:403–412

Nolte MA, Arens R, Kraus M, van Oers MH, Kraal G, van Lier RA, Mebius RE (2004) B cells are crucial for both development and maintenance of the splenic marginal zone. J Immunol 172:3620–3627

Nolte MA, Belien JA, Schadee-Eestermans I, Jansen W, Unger WW, van Rooijen N, Kraal G, Mebius RE (2003) A conduit system distributes chemokines and small blood-borne molecules through the splenic white pulp. J Exp Med 198:505–512

Ohno S (1970) Evolution by gene duplication. Springer-Verlag, New York

Oliver AM, Martin F, Gartland GL, Carter RH, Kearney JF (1997) Marginal zone B cells exhibit unique activation, proliferative and immunoglobulin secretory responses. Eur J Immunol 27:2366–2374

Oliver AM, Martin F, Kearney JF (1999) IgMhighCD21high lymphocytes enriched in the splenic marginal zone generate effector cells more rapidly than the bulk of follicular B cells. J Immunol 162:7198–7207

Putnam NH, Butts T, Ferrier DE, Furlong RF, Hellsten U, Kawashima T, Robinson-Rechavi M, Shoguchi E, Terry A, Yu JK, Benito-Gutierrez EL, Dubchak I, Garcia-Fernandez J, Gibson-Brown JJ, Grigoriev IV, Horton AC, de Jong PJ, Jurka J, Kapitonov VV, Kohara Y, Kuroki Y, Lindquist E, Lucas S, Osoegawa K, Pennacchio LA, Salamov AA, Satou Y, Sauka-Spengler T, Schmutz J, Shin IT, Toyoda A, Bronner-Fraser M, Fujiyama A, Holland LZ, Holland PW, Satoh N, Rokhsar DS (2008) The amphioxus genome and the evolution of the chordate karyotype. Nature 453:1064–1071

Reif K, Ekland EH, Ohl L, Nakano H, Lipp M, Forster R, Cyster JG (2002) Balanced responsiveness to chemoattractants from adjacent zones determines B-cell position. Nature 416:94–99

Rogozin IB, Iyer LM, Liang L, Glazko GV, Liston VG, Pavlov YI, Aravind L, Pancer Z (2007) Evolution and diversification of lamprey antigen receptors: evidence for involvement of an AID-APOBEC family cytosine deaminase. Nat Immunol 8:647–656

Rubtsov AV, Swanson CL, Troy S, Strauch P, Pelanda R, Torres RM (2008) TLR agonists promote marginal zone B cell activation and facilitate T-dependent IgM responses. J Immunol 180:3882–3888

Scandella E, Bolinger B, Lattmann E, Miller S, Favre S, Littman DR, Finke D, Luther SA, Junt T, Ludewig B (2008) Restoration of lymphoid organ integrity through the interaction of lymphoid tissue-inducer cells with stroma of the T cell zone. Nat Immunol 9:667–675

Shiow LR, Rosen DB, Brdickova N, Xu Y, An J, Lanier LL, Cyster JG, Matloubian M (2006) CD69 acts downstream of interferon-alpha/beta to inhibit S1P1 and lymphocyte egress from lymphoid organs. Nature 440:540–544

Slifka MK, Antia R, Whitmire JK, Ahmed R (1998) Humoral immunity due to long-lived plasma cells. Immunity 8:363–372

Sun Z, Unutmaz D, Zou YR, Sunshine MJ, Pierani A, Brenner-Morton S, Mebius RE, Littman DR (2000) Requirement for RORgamma in thymocyte survival and lymphoid organ development. Science 288:2369–2373

Ueno T, Hara K, Willis MS, Malin MA, Hopken UE, Gray DH, Matsushima K, Lipp M, Springer TA, Boyd RL, Yoshie O, Takahama Y (2002) Role for CCR7 ligands in the emigration of newly generated T lymphocytes from the neonatal thymus. Immunity 16:205–218

Vondenhoff MF, Desanti GE, Cupedo T, Bertrand JY, Cumano A, Kraal G, Mebius RE, Golub R (2008) Separation of splenic red and white pulp occurs before birth in a LTalpha-beta-independent manner. J Leukoc Biol 84:152–161

Watts TH (2005) TNF/TNFR family members in costimulation of T cell responses. Annu Rev Immunol 23:23–68

Wilson M, Hsu E, Marcuz A, Courtet M, Du Pasquier L, Steinberg C (1992) What limits affinity maturation of antibodies in Xenopus – the rate of somatic mutation or the ability to select mutants? Embo J 11:4337–4347

Withers DR, Kim MY, Bekiaris V, Rossi SW, Jenkinson WE, Gaspal F, McConnell F, Caamano JH, Anderson G, Lane PJ (2007) The role of lymphoid tissue inducer cells in splenic white pulp development. Eur J Immunol 37:3240–3245

Woolf E, Grigorova I, Sagiv A, Grabovsky V, Feigelson SW, Shulman Z, Hartmann T, Sixt M, Cyster JG, Alon R (2007) Lymph node chemokines promote sustained T lymphocyte motility without triggering stable integrin adhesiveness in the absence of shear forces. Nat Immunol 8:1076–1085

Zapata A, Ameimiya CT (2000) Phylogeny of lower vertebrates and their immunological structures. In: du Pasquier L, Litman GW (eds) Origin and evolution of the vertebrate immune system. Springer, Berlin, pp 67–110

# Part V
# Disassembling the Puzzle: Effect of Aging

# Chapter 13
# Age-Associated Decline in Peripheral Lymphoid Organ Functions

**Rania M. El Sayed, John G. Tew, and Andras K. Szakal**

**Abstract** Immunosenescence occurs after maturation and is expressed as a series of failing immunologic functions. Aging is best studied in the peripheral organs of the immune system such as lymph nodes, spleen, and isolated lymphoid tissues. The functional immunological building blocks of peripheral lymphoid organs are the lymphoid follicles. Aging defects occur in the components of lymphoid follicles, such as the T cells, B cells, and accessory cells, e.g., the dendritic cells. Among the dendritic cells, the most influential on the aging humoral immune response is the follicular dendritic cell (FDC). In aged mice, the Ab response is depressed because of the reduced capacity of FDCs to trap ICs needed for stimulation of B cells. This path is interrupted by the 70% age-related reduction of Fc$\gamma$RII on FDCs and the 88% age-related reduction in IC trapping. Thus, we conclude that these reductions account, in large part, for the declining Ab response in aged animals.

## 13.1 Introduction

In man and other mammals, efficient protection against invading pathogens requires the presence of highly compartmentalized peripheral lymphoid organs. These highly compartmentalized peripheral lymphoid organs are the anatomical sites for the initiation of specific immune responses. Such organs include the lymph nodes and the spleen and simple lymphoid tissues, such as solitary intestinal and respiratory lymphoid nodules associated with the mucosa and the Peyer's patches of the ileum. The functional building block in these tissues and organs is the lymphoid

R.M.E. Sayed and J.G. Tew
Department of Microbiology and Immunology, Virginia Commonwealth University, Richmond, VA 23298-0678, USA

A.K. Szakal (✉)
Department of Anatomy and Neurobiology, and The Immunobiology Group, Virginia Commonwealth University, Richmond, VA 23298-0678, USA
e-mail: aszakal@comcast.net

P. Balogh (ed.), *Developmental Biology of Peripheral Lymphoid Organs*,
DOI 10.1007/978-3-642-14429-5_13, 

nodule. Lymphoid nodules contain the required cell types for the humoral immune response, i.e., lymphocytes (T and B cells), dendritic cells (DCs), and follicular dendritic cells (FDCs). Upon immunization, lymphoid nodules develop germinal centers (GCs). Aging affects cells in the lympoid nodules to varying degrees and at different times. However, due to defects with FDC functions, B and T cells do not function optimally to form GCs. GCs are markedly reduced in both size and numbers and we believe that problems associated with aged FDCs may explain an age-associated decline in peripheral lymphoid organ function and, thus, in the humoral immune response.

## 13.2 The Aging of the Immune System

During aging, a variety of complex and widespread alterations of the immune system have been well documented, which culminate in immunosenescence. Deleterious changes may extend throughout the immune system from hematopoietic stem cells and lymphoid progenitors to mature lymphocytes and secondary lymphoid organs (Kovaiou et al. 2007). For example, the breakdown of epithelial barriers (Gomez et al. 2005), impaired neutrophil production of superoxide anions, decreased chemotaxis (Fulop et al. 2004), reduced macrophage respiratory burst, and diminished antigen (Ag) presentation (Davila et al. 1990; Villanueva et al. 1990; Kissin et al. 1997; Tasat et al. 2003) have been found in aged immune systems. In addition, reduced expression of Toll-like receptors (Plowden et al. 2004), impaired natural killer (NK) cell cytotoxicity, and decreased production of cytokines (Mocchegiani et al. 2003; Mariani et al. 2002) were associated with age-related changes of the immune system.

At the level of the adaptive immune system, age-related defects in T cells (Taub and Longo 2005; Haynes et al. 1999; Appay et al. 2002; Grubeck-Loebenstein and Wick 2002; Pawelec et al. 2002; Haynes et al. 2003; Naylor et al. 2005; Haynes and Swain 2006; Haynes et al. 2005) and, to some extent, in B cells (Schwab et al. 1989; Miller and Kelsoe 1995; Weksler and Szabo 2000; Johnson and Cambier 2004; Chong et al. 2005) have been described. Further, with aging, a paucity of GCs (Hanna et al. 1967) and a depressed secondary humoral immune response (Legge and Austin 1968) have been reported. GCs are known to be the sites of B memory cell development (Thorbecke et al. 1974; Klaus et al. 1980; Klaus and Kunkl 1982; Coico et al. 1983). B memory cell development is dependent on antigen trapping in lymphoid nodules (White et al. 1970; Thorbecke et al. 1974; Thorbecke and Lerman 1976). Antigen trapping in lympoid nodules is, in turn, dependent on the antigen transport mechanism and the trapping of Ag by FDCs, which form the antigen retaining reticulum or network (Nossal et al. 1968; Szakal and Hanna 1968; Hanna and Szakal 1968; Szakal et al. 1983).

FDCs function as members of the "alternative antigen transport pathway." This pathway, consisting of the antigen transport cell–Immune Complex Coated body-B cell axis (ATC–FDC–ICCOSOME), leads to the formation of GCs where antibody-forming cell and memory B cell development are initiated through the interaction of FDC-retained antigen, B cells, and T helper cells. Evidence suggests that these interactions are downregulated through antibody feedback or, as needed, reactivated with the utilization of FDC-retained antigen for the maintenance of antibody levels. Age-related FDC defects can seriously compromise the capacity of this pathway to maintain immunity. Thus, as integral parts of the microenvironment of the follicle, FDCs play a pivotal role in the initiation and maintenance of the secondary antibody response (Szakal et al. 1989; Tew et al. 1989; Tew et al. 2001).

## 13.3 The FDCs Relationship to Other Dendritic Cells

The pivotal role of the FDC in the initiation and maintenance of the secondary antibody response renders its relationship to other DCs important. FDCs are located in the B cell regions of secondary lymphoid tissues, i.e., in the light zones of GCs in the lymphoid follicles (Szakal and Hanna 1968; Tew et al. 2001; Trends in Immunology 2001). FDCs function in the binding and retention of immune complexes and possess potent accessory cell functions related to their interaction with B cells (Tew et al. 1979; Schnizlein et al. 1984; Tew et al. 1990). In contrast, T cell-associated DCs are located in the thymus-dependent regions of secondary lymphoid tissues where they have certain accessory functions for T cells. These DCs can also be found performing accessory functions for T cells as the interdigitating cells of the thymic medulla, the lymph node paracortex and the splenic periarteriolar lymphocytic sheath (PALS) (see review by Kohler et al. 1990). In addition, DCs are found in epithelia and various connective tissues (e.g., epidermal and mucosal Langerhan cells), serving similar T cell-related accessory functions. Therefore, the FDC is a dendritic cell type; however, the FDC is morphologically and functionally distinct from the T cell-associated dendritic cell, known as the DC.

## 13.4 The Molecular Model of FDC–B Cell Interactions

In 2001, we described the ligand–receptor interactions between FDC and B cells that appear to signal and stimulate B cells in a manner that "goes beyond the necessity of T cell help" (Tew et al. 2001).

In summary, this model for FDC/B cell interaction was explained and illustrated as (1) the trapping of Ag–antibody (Ab) complexes by FDCs; (2) the presentation by FDCs of Ag to B cell receptors (BCR) of germinal center B cells; (3) the provision by the Ag–BCR interaction of a positive signal for B cell activation and differentiation; (4) the presentation by FDCs of complement-derived CD21 ligand (CD21L) to B cell CD21; (5) the interaction of the FDC-CD21L with the CD21–CD19–CD81 complex; and (6) through the delivery of a positive cosignal for B cell activation and differentiation (Tew et al. 2001).

In Fig. 13.1a, b, we have distinguished this molecular model for young and old animals. The molecular interactions in young animals between FDCs and immune complexes (ICs), which are the combined form of Ag and Ab, and the interactions between FDCs and B cells via the ICs and bound complement fragments of C3 (iC3b, C3dg, and C3d) that represent the CD21 ligand (CD21L) are illustrated in Fig. 13.1a.

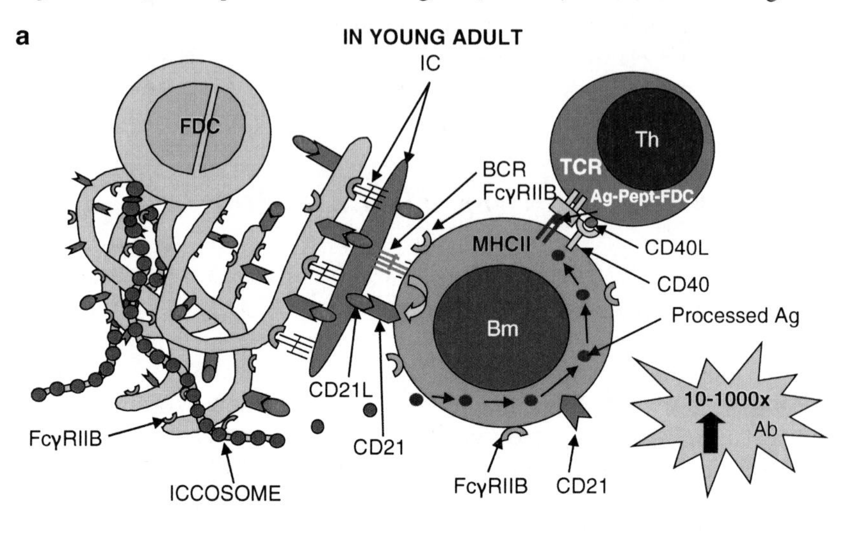


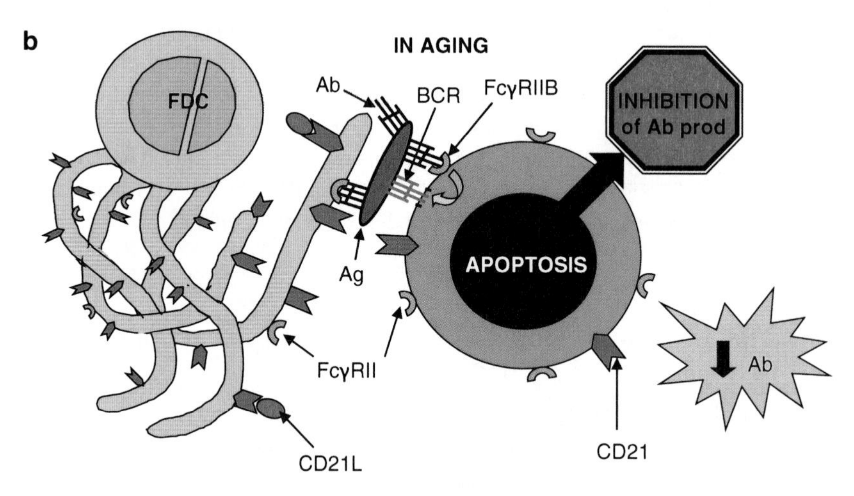


**Fig. 13.1** (**a**) Young FDCs are illustrated, based on histochemical evidence, with numerous surface molecules along with a B cell and a T helper cell. Note the presence of numerous CD21 molecules on FDCs decorated with CD21L (L, ligands) and numerous FcγRII on the dendrites. The antibodies (Ab) and the antigen (Ag) form the immune complex and with the aid of the IC bind the FDC to the B cell. The B cell endocytose the ICCOSOMEs, process the FDC-derived Ag and present it to the T helper cell attached to the B cell. (**b**) Old FDCs, based on histochemical evidence, are Illustrated with few FcγRIIB to bind Abs and few CD21L to bind Ag and B cells. No ICCOSOMEs are produced. The IC cross-links the BCR and the FcγRII, which results in apoptosis of the B cell and, consequently, in a lack of Ab production

In young animals, FDCs trap ICs and provide intact Ag for interaction with BCR on GC B cells. This provides a positive signal for B cell activation and differentiation. The FDC–CD21L binds complement receptor 2 (CR2 or CD21) on the B cell via CD21 in the B cell coreceptor complex consisting of CD19/CD21/TAPA-1 (Faeron and Carter 1995). Engagement of CD21 in the B cell coreceptor complex by complement-derived FDC–CD21L delivers a critical cosignal. Coligation of BCR and CD21 with a single molecule of Ag facilitates association of the two receptors, and the cytoplasmic tail of CD19 is phosphorylated by a tyrosine kinase associated with the B cell receptor complex (Carter et al. 1997). The curved arrow pointing from the BCR to CD21 in Fig. 13.1 is intended to indicate this known interaction. This cosignal dramatically augments stimulation delivered by engagement of BCR by Ag. Conversely, a blockade of FDC–CD21L reduces the immune responses considerably (typically 10–1,000-fold) (Qin et al. 1998).

Immunogens are almost instantaneously converted into ICs by Ab persisting in immune animals from prior immunizations, and ICs form in primary responses as soon as the first Ab is produced. These ICs are trapped by FDCs and this leads to GC formation. The FDC-trapped ICs are poorly immunogenic in vitro, yet a minimal amount of Ag converted into FDC-trapped ICs in vivo provokes powerful recall responses. Our results indicate that FDCs render ICs highly immunogenic. In fact, in the presence of FDCs, the ICs are more immunogenic than free Ag (Tew et al. 2001).

A high density of Fc$\gamma$RIIB on young adult FDCs binds Ab in the IC. Consequently, the immunoreceptor tyrosine-based inhibition motif (ITIM) signal delivered via B cell Fc$\gamma$RIIB may be blocked. This inhibitory signal is initiated by Ag–Ab complexes cross-linking BCR and Fc$\gamma$RIIB on B cells, as shown in Fig. 13.1a. Note that BCR is not cross-linked with B cell Fc$\gamma$RIIB in the model because a high concentration of Fc$\gamma$RIIB on FDCs minimizes a negative signal to the B cell (Tew et al. 2001).

Further, FDCs provide IC-coated bodies (ICCOSOMES), which B cells find highly palatable (Szakal et al. 1988). The iccosomal membrane is derived from FDC membranes that have Ag, CD21L, and Ig-Fc attached. Iccosomes bind tightly to B cells and are rapidly endocytosed (Szakal et al. 1988). The binding of BCR and CD21 of the B cell to the iccosomal Ag–CD21L–Ig-Fc complex is postulated as being critical to the endocytic process. The B cell processes the FDC-derived Ag, which is presented to T cells to obtain T cell help (Kosco et al. 1988). Thus, FDC-facilitated ligand–receptor interactions help stimulate B cells and provide assistance to B cells beyond that provided by T cells.

## 13.5 FDCs in Aging

A series of studies were conducted to examine the age-related defects in FDC functions (reviewed in Szakal et al. 2002). In one study, the FDC reactive mAb, FDC-M1, was used (Aydar et al. 2000) to assess the development of FDC-network and to determine the size of FDC-reticula in OVA-immunized young and old mice. FDC-M1 labeling showed surprisingly well-developed FDC-networks in old mice.

FDC-M1 positive FDC-reticula enlarged gradually and peaked at Day 5 in both young and old mice. Immunohistochemical and flow cytometric studies showed that both the average density of FDC-M1 labeling and FDC-reticulum size were similar in young and old mice. It was concluded that aging does not affect the development and kinetics of FDC-network, and that reduced IC trapping in old mice is not due to deficient FDC numbers in the reticula (Aydar et al. 2000).

The reactivity of FDC-M2 with FDCs is dependent on trapped ICs capable of fixing complement. FDC-M2 reacts specifically with the C4 fragment of complement associated with the FDC-ICs (Taylor et al. 2002). Studies of the kinetics of FDC-M2 labeling were used as a functional parameter reflecting antigen localization. Correlating this with the presence of FDC-FcγRII (Fig. 13.2a, b) and complement receptors showed an impairment in the functional capacity of aged FDCs to trap and retain ICs (Aydar et al. 2004b). In these studies, the size of FDC-M2 positive FDC-reticula in young mice increased gradually and peaked at Day 5, comparable to FDC-Ml and CR1 and 2 labeling. In old mice, however, the size of FDC-M2 positive FDC-reticula decreased by 88%, suggesting impairment in the capacity of aging FDC-reticulum to retain ICs (note the reduced number of ICs illustrated in Figs. 13.1b and 13.2a, b, despite the maintenance of its structural integrity). The same study showed only an insignificant reduction in the average size of CR1 and 2 positive FDC-reticula in old mice. The difference between the CRI and 2 positive FDC-reticulum-size of young and old mice during the GC response was also insignificant (Aydar et al. 2004a).

Anti-FcγR monoclonal antibody (clone 2.4G2), which recognizes both mouse FcγRII and FcγRIII, was used to study FcγRII expression in old mice (Aydar et al. 2004a). The average density of FcγRII labeling on old FDCs was dramatically reduced by >70% in the 3–10-day period of the GC reaction (as illustrated in Figs. 13.1b and 13.2a). The density of FDC-network numbers and sizes was also radically diminished by >90%. This reduction indicated that not all of the FDCs in an FDC-M1 positive FDC-network express substantial levels of FcγRII, and that in old mice, only some reticula have FDCs expressing FcγRII. This observation correlated with the reduction in IC trapping by old FDCs in the original antigen localization studies and was confirmed with FDC-M2 labeling (for IC binding) in old mice. Moreover, reduced PNA positive GC numbers in old mice correlated well with the reduced FDC-M2 staining (Coico et al. 1983).

In addition, FDC-FcγRIIB binds the Fc regions of immunoglobulins thereby inhibiting the SH2 domain-containing inositol phosphatase (SHIP) signaling pathway and minimizing ITIM activation in B cells by reducing co-cross-linking of BCR and B cell FcγRIIB (Qin et al. 2000; Aydar et al. 2003, 2004b). Therefore, it is possible that the lack of FcγRII on old FDCs would correlate with ITIM activation as indicated by experiments with SHIP phosphorylation. This hypothesis was tested using anti-BCR-stimulated A20 cells (Aydar et al. 2004b). In this in vitro experiment in the presence of young FDC, reduced ITIM signaling occurred, as indicated by a 60% reduction in SHIP phosphorylation. On the other hand, incubation of A20 cells with old FDC, or FDC with blocked FcγRII, failed to block SHIP phosphorylation in A20 cells stimulated with anti-BCR. As expected, young FDC blocked

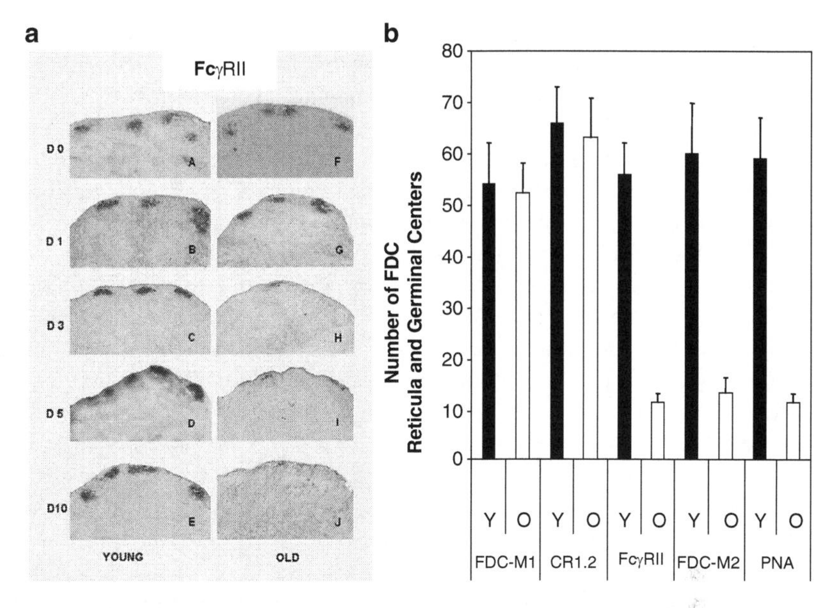


**Fig. 13.2** (**a**) This shows a series of representative light micrographs illustrating the kinetics of immunohistochemical labeling with FDC-M2 that binds to immune complexes in FDC reticula of draining axillary lymphnodes of actively immunized young and old mice. Note the striking decrease over time of labeling (of ICs) of old FDC reticula (D, day). The same relationships existed in brachial and popliteal lymph nodes. (**b**) This shows the correlation between the number of FDC reticula expressing FcγRII, IC trapping (indicated by FDC-M2), and a reduced number of PNA-positive germinal centers in old mice. Young (Y) and old (O) represent results from mice actively immunized and challenged with OVA in each front and hind legs. The data represents the mean number of FDC reticula positive for FDC-M1, anti-CR1/2, anti-FcγRII (FcγRII), or the mean number of germinal centers positive for PNA±SEM. Significant differences are indicated by the *p* values. These data were obtained from mice taken at 5 days but the same relationships exist at day 3 and day 10. Diagrams taken from reference Aydar et al. (2003)

SHIP phosphorylation; however, old FDC did not and were like control FDC that lacked FcγRII. The FDC-FcγRII-mediated rescue of B cells deteriorates with aging leading to B cell inactivation and significantly depressed secondary antibody responses (Fig. 13.1b). This was confirmed by further studies in which old FDCs cultured with IC-stimulated OVA-immune-FcγRIIB$^{-/-}$ lymphocytes showed an increase in antibody production to near normal levels (Aydar et al. 2004a).

## 13.6 Antigen Transport and Reticula in Old Mice

Studies using old C57BL/6 and BALB/c mice revealed that antigen transport is incomplete and that only a small fraction of antigen transport sites develop (Szakal et al. 1983; Holmes et al. 1984; Szakal et al. 1990). In old mice, antigen transporting cells (ATCs) show a tendency to accumulate in the subcapsular sinus. Only some of these cells transport small amounts of Ag–Ab complexes to the follicles. Other ATCs appear to lack the capacity to transport ICs on their surfaces and thus join the FDC-network without any trapped Ag–Ab complexes (Tew et al. 1997).

The deficit in antigen transport corresponded to a similar deficit in the number of FDC retaining reticula by Day 1–3 after antigen challenge (Kosco et al. 1989). Many old ATCs failed to develop dendrites and some old FDCs were ultrastructurally atrophic, retained little antigen, and produced no iccosomes. Nevertheless, the normal 1:1 reduced FDC-reticulum to reduced GC ratio, an expression of the antigen retention requirement for GC development, still existed in old mice (White et al. 1970; Kosco et al. 1989; Szakal et al. 1992); however, due to antigen transport defects, significantly ($\sim$90%) fewer FDC-reticula developed, and, in turn, proportionally fewer GCs were induced (Szakal et al. 1990).

Since GCs are the sites where early antibody forming cells (Tew et al. 1980; Szakal et al. 1983; Tew et al. 1989; Kosco et al. 1989) and, subsequently, memory B cells develop, these results support the hypothesis that a reduced capacity for antigen transport is responsible, at least in part, for the depressed secondary humoral immune response in old mice (Coico et al. 1983; Kosco et al. 1988).

The reduced levels of FcγRII on old FDCs would implicate a defect in a pathway leading to synthesis or trafficking of synthesized FcγRII on old FDC surfaces. It is plausible then that old FDCs, unlike young FDCs, would not respond to external stimuli, such as IC trapping via FcγRIIB and/or TLR4 ligation, by upregulating the surface expression of important molecules such as FDC-ICAM-1, FDC-VCAM-1, and FDC-FcγRIIB (El Shikh et al. 2006). Additional data clarified that expression of FDC molecules that probably function in FDC–B cell interactions is subject to regulation and that agents known to activate other cells will increase some FDC molecules (El Shikh et al. 2009).

Specifically, binding of ICs through FcγRIIB resulted in dramatic upregulation of ICAM-1, VCAM-1, and FcγRIIB on FDCs (El Shikh et al. 2006). Moreover, our studies provided evidence for the involvement of FcγRIIB-IC-mediated activation of FDCs in the induction of FDC-BAFF (El Shikh et al. 2009). It is thus reasonable that the inability of old FDCs to upregulate the surface expression of FcγRIIB would render them nonresponsive to stimuli involving IC retention, which would lead to impaired production of soluble factors important in GC-development and the maintenance of Ab production.

## 13.7 Old B and T Cells

Experiments conducted to determine the contribution of deficits in aged B and T cells to the poor immune responses in the aged showed no significant differences between young and old B cells in cocultures with young FDCs. The same study also concluded that the age of B-lymphocytes does not modify the accessory activity of FDCs and that there was no difference in antibody production between cocultures using young or old T cells. Old T cells did not seem to modify the accessory activity of young FDCs even at 1/16th of the concentration they were present in the old lymph node (Aydar et al. 2000; Szakal et al. 2002). In this study, antigen-primed mice were used to isolate 90% pure populations of T and B cells using negative selection. The T and B cells were mixed for the cultures according to B and T cell ratios determined for young and old lymph nodes by flow cytometric analysis. The numbers of T cells were subsequently reduced logarithmically in the cell population to 1/16th of the original ratio. Therefore, this study concluded that the defects in the GC reaction and the anamnestic antibody response were due to the defect in FDCs rather than in B or T cells (Aydar et al. 2000; Szakal et al. 2002).

The ability to mend the defects in the immune response via the repair of the altered activities in aged FDC-delivered signals further supports the role of FDCs rather than B or T cells in the age-related impairment of the immune response. In a study aimed at analyzing this concept, young and old FDCs were cocultured with LPS-stimulated B lymphocytes in the presence of ICs with active or inactivated complement. The reason for adding ICs and complement was that the new ICs would bind to the FDCs, activate the complement cascade, and bind more CD21L. These ligands were then expected to bind the FDCs to the B cells for costimulation. The results showed that cocultures with old FDCs that were supplemented with ICs and active complement produced nearly as much antibody as cocultures with young FDCs (Aydar et al. 2000; Szakal et al. 2002).

Multiple B cell defects are still described to explain the dysfunctional immune system in the elderly [reviewed in (Colonna-Romano et al. 2008)]. Repeated intermittent or chronic antigen exposure may lead to lymphocyte clonal exhaustion thus compromising immunity in the elderly, who have experienced extensive exposure to infectious agents, auto-antigens, and cancer antigens. Age is a condition characterized by lack of B cell clonotypic immune response to new extracellular pathogens, and data suggest that the loss of naïve B cells could represent a hallmark of immunosenescence (Colonna-Romano et al. 2008). In addition, reduced numbers of total B lymphocytes, shrinkage in receptor repertoire, increased amounts of auto-antibodies, reduced formation of GCs, and impaired generations of high-affinity antibodies all have been reported in old ages and centenarians (Colonna-Romano et al. 2008).

Moreover, it has also been shown that resting aged B cells exhibited similar levels of CD40 expression when compared with young cells and efficiently upregulated CD86 and CD69 and downregulated CD38 upon stimulation both with LPS and CD4O. However, aged B cells proliferated less than young B cells and showed a consistent, but not statistically significant, reduction in their ability to form blast cells. Importantly, in all cases, stimulated aged B cells always showed a greater response than unstimulated cells, suggesting that certain functions were retained (Blaeser et al. 2008). The same group also reported that aged B cells were capable of undergoing normal activation events in vivo including germinal center reactions, but were defective in expansion following T cell-dependent stimulation (Dailey et al. 2001). Supporting a role for the microenvironment in these age-related dysfunctions, Song et al. (1997) reported that aged splenocytes appear to be comparable in expression and function of CD40 to their young counterparts and that the proportion of CD40 positive resting B cells remains relatively stable during aging (Yang et al. 1996; Song et al. 1997).

Consistent with the studies of Szakal and coworkers (Szakal et al. 2002; Aydar et al. 2004a) on FDCs and germinal center reactions in aging, Cozine et al. (2005) and McGlauchlen and Vogel (2003) agreed that the defective ability of FDCs to trap ICs, the lower levels of CD21L on FDC surfaces, and the reduced expression of FDC-FcγRIIB in aged mice account for diminished GC responses. Moreover, Hakim and Gress (2007) also attributed the reduced induction of T cell-dependent B cell responses and the limited persistence of antigen-specific memory B cell populations in the elderly to mechanisms involving FDC dysfunctions.

Studies of the role of FDC-FcγRIIB in the activation of FDC by ICs (El Shikh et al. 2006) have clarified that expression of FDC accessory molecules is subject to regulation, and that ICs, TLR ligands (Pawelec et al. 2002), and the binding to collagen type 1 (El Shikh et al. 2007) induced FDC activation and upregulation of FDC accessory molecules. Furthermore, addition of ICs, in vitro, to purified FDCs from wild-type mice, but not from FcγRIIB$^{-/-}$ mice, prompted the production of mRNA for FcγRIIB, ICAM-1, and VCAM-1 within 3 h, indicating that engagement of ICs to FDCs is important in initiating pathways leading to FDC activation (El Shikh et al. 2006).

## 13.8 Summary and Conclusion

The reduced numbers of GCs accompanying aging correlates well with the impaired immune responses to recall Ags. GCs are the sites where FDCs orchestrate, together with B and T cells, the development of an immune response towards antigenic challenge. Efficient GC reactions result in the development of memory B cells and high affinity-Ab-secreting plasma cells. Expression of FDC-FcγRII is dramatically reduced in GC reactions in old mice and the reduction correlates with reduced Ag–Ab retention. FDCs from FcγRII$^{-/-}$ mice do not retain and present ICs to wild-type B cells effectively either in vivo or in vitro. Germinal center B cells rapidly proliferate in response to antigenic challenge in young mice; however, in old mice the GC reaction is limited. Since aged FDCs lose their capacity to retain ICs and present Ag to B cells, it was not surprising that their GC reactions and antibody responses were reduced. Thus, FDCs appear to represent a "major bottle neck" in the induction of antibody responses in aged animals (Aydar et al. 2004a).

## References

Appay VP, Dunbar R, Callan M, Klenerman P, Gillespie GM, Papagno L, Ogg GS, King A, Lechner F, Spina CA, Little S, Havlir DV, Richman DD, Gruener N, Pape G, Waters A, Easterbrook P, Salio M, Cerundolo V, McMichael AJ, Rowland-Jones SL (2002) Memory CD8+ T cells vary in differentiation phenotype in different persistent virus infections. Nat Med 8:379–385

Aydar Y, Balogh P, Tew JG, Szakal AK (2000) Age-related depression of FDC accessory functions and CD21 ligand-mediated repair of co-stimulation. Eur J Immunol 32:2817–2826

Aydar Y, Balogh P, Tew JG, Szakal AK (2003) Altered regulation of Fc gamma RII on aged follicular dendritic cells correlates with immunoreceptor tyrosine-based inhibition motif signaling in B cells and reduced germinal center formation. J Immunol 171: 5975–5987

Aydar Y, Balogh P, Tew JG, Szakal AK (2004a) Follicular dendritic cells in aging, a "bottle-neck" in the humoral immune response. Ageing Res Rev 3:15–29

Aydar Y, Wu J, Song J, Szakal AK, Tew JG (2004b) FcγammaRII expression on follicular dendritic cells and immunoreceptor tyrosine-based inhibition motif signaling in B cells. Eur J Immunol 34:98–107

Blaeser A, McGlauchlen K, Vogel LA (2008) Aged B lymphocytes retain their ability to express surface markers but are dysfunctional in their proliferative capability during early activation events. Immun Ageing 5:15

Carter RH, Doody GM, Bolen JB, Fearon DT (1997) Membrane IgM-induced tyrosine phosphorylation of CD19 requires a CD19 domain that mediates association with components of the B cell antigen receptor complex. J Immunol 158:3062–3069

Chong Y, Ikematsu H, Yamaji K, Nishimura M, Nabeshima S, Kashiwagi S, Hayashi J (2005) CD27(+) (memory) B cell decrease and apoptosis-resistant CD27(-) (naive) B cell increase in aged humans: complications for age-related peripheral B cell developmental disturbances. Int Immunol 17:383–390

Coico RF, Bhogal BS, Thorbecke GJ (1983) Relationship of germinal centers in lymphoid tissue to immunologic memory. VI. Transfer of B cell memory with lymph node cells fractionated according to their receptors for peanut agglutinin. J Immunol 131(5):2254–2257

Colonna-Romano G, Bulati M, Aquino A, Vitello S, Lio D, Candore G, Carus C (2008) B cell immunosenescence in the elderly and in centenarians. Rejuvenation Res 11:433–439

Cozine CL, Wolniak KL, Waldschmidt TJ (2005) The primary germinal center response in mice. Curr Opin Immunol 17:298–302

Davila DR, Edwards CK, III, Arkins S, Simon J, Kelley KW (1990) Interferon-gamma-induced priming for secretion of superoxide anion and tumor necrosis factor-alpha declines in macrophages from aged rats. FASEB J 4:2906–2911

Dailey RW, Eun SY, Russell CE, Vogel LA (2001) B cells of aged mice show decreased expansion in response to antigen, but are normal in effector function. Cell Immunol 214:99–109

El Shikh ME, El Sayed R, Szakal AK, Tew JG (2006) Follicular dendritic cell (FDC)-FcgammaR-IIB engagement via immune complexes induces the activated FDC phenotype associated with secondary follicle development. Eur J Immunol 36:2715–2724

El Shikh ME, El Sayed RM, Tew JG, Szakal AK (2007) Follicular dendritic cells stimulated by collagen type I develop dendrites and networks in vitro. Cell Tissue Res 329:81–89

El Shikh ME, El Sayed RM, Szakal AK, Tew JG (2009) T-independent antibody responses to T-dependent antigens: a novel follicular dendritic cell-dependent activity. J Immunol 182 (6):3482–3491

Faeron DT, Carter RH (1995) The CD19-CR2-TAPA-1 complex of B lymphocytes: linking natural to acquired immunity. Ann Rev Immunol 13:127–149

Fulop T, Larbi A, Douziech N, Fortin C, Guerard KP, Lesur O, Khali Al, Dupuis G (2004) Signal transduction and functional changes in neutrophils with aging. Aging Cell 3:217–226

Gomez CR, Boehmer ED, Kovacs EJ (2005) The aging innate immune system. Curr Opin Immunol 17:457–462

Grubeck-Loebenstein B, Wick G (2002) The aging of the immune system. Adv Immunol 80:243–284

Hakim FT, Gress RE (2007) Immunosenescence: deficits in adaptive immunity in the elderly. Tissue Antigens 70:179–189

Hanna MG Jr, Szakal AK (1968) Localization of 125I-labeled antigen in germinal centers of mouse spleen: histologic and ultrastructural autoradiographic studies of the secondary immune reaction. J Immunol 101:949–962

Hanna MG Jr, Nettesheim P, Ogden L, Makinodan T (1967) Reduced immune potential of aged mice: significance of morphologic changes in lymphatic tissue. Proc Soc Exp Biol Med 125:882–886
Haynes L, Linton PJ, Eaton SM, Tonkonogy SL, Swain SL (1999) Interleukin 2, but not other common gamma chain-binding cytokines, can reverse the defect in generation of CD4 effector T cells from naive T cells of aged mice. J Exp Med 190:1013–1024
Haynes L, Swain SL (2006) Why aging T cells fail: implications for vaccination. Immunity 24:663–666
Haynes L, Eaton SM, Burns EM, Randall TD, Swain SL (2003) CD4 T cell memory derived from young naive cells functions well into old age, but memory generated from aged naive cells functions poorly. Proc Natl Acad Sci USA 100:15053–15058
Haynes L, Eaton SM, Burns EM, Randall TD, Swain SL (2005) Newly generated CD4 T cells in aged animals do not exhibit age-related defects in response to antigen. J Exp Med 201:845–851
Holmes KL, Schnizlein CT, Perkins EH, Tew JG (1984) The effect of age on antigen retention in lymphoid follicles and in collagenous tissue of mice. Mech Ageing Dev 25:243–255
Johnson SA, Cambier JC (2004) Ageing, autoimmunity and arthritis: senescence of the B cell compartment – implications for humoral immunity. Arthritis Res Ther 6:131–139
Kissin E, Tomasi M, McCartney-Francis N., Gibbs CL, Smith PD (1997) Age-related decline in murine macrophage production of nitric oxide. J Infect Dis 175:1004–1007
Klaus GG, Humphrey JH, Kunkl A, Dongworth DW (1980) The follicular dendritic cell: its role in antigen presentation in the generation of immunological memory. Immunol Rev 53:3–28. Review
Klaus GG, Kunkl A (1982) The role of T cells in B cell priming and germinal centre development. Adv Exp Med Biol 149:743–749
Kohler H, Bona C (eds); Steiman RM (guest ed) (1990) Dendritic sells. Int Rev Immunol 6(2–6):89–206
Kosco MH, Burton GF, Kapasi ZF, Szakal AK, Tew JG (1989) Antibody-forming cell induction during an early phase of germinal centre development and its delay with ageing. Immunology 68:312–318
Kosco MH, Szakal AK, Tew JG (1988) In vivo obtained antigen presented by germinal center B cells to T cells in vitro. J Immunol 140:354–360
Kovaiou RD, Herndler-Brandstetter D, Grubeck-Loebenstein B (2007) Age-related changes in immunity: implications for vaccination in the elderly. Expert Rev Mol Med 9:1–17
Legge JS, Austin CM (1968) Antigen localization and the immune response as a function of age. Aust J Exp Biol Med Sci 46:361–365
Mariani E, Meneghetti A, Neri S, Ravaglia G, Forti P, Cattini L, Facchini A (2002) Chemokine production by natural killer cells from nonagenarians. Eur J Immunol 32:1524–1529
Mocchegiani E, Muzzioli M, Giacconi R, Cipriano C, Gasparini N, Franceschi C, Gaetti R, Cavalieri E, Suzuki H (2003) Metallothioneins/PARP-1/IL-6 interplay on natural killer cell activity in elderly: parallelism with nonagenarians and old infected humans. Effect of zinc supply. Mech Ageing Dev 124:459–468
Miller C, Kelsoe G (1995) Ig VH hypermutation is absent in the germinal centers of aged mice. J.Immunol 155:3377–3384
McGlauchlen KS, Vogel LA (2003) Ineffective humoral immunity in the elderly. Microbes Infect 5:1279–1284
Naylor K, Li G, Vallejo AN, Lee WW, Koetz K, Bryl E, Witkowski J, Fulbright J, Weyand CM, Goronzy JJ (2005) The influence of age on T cell generation and TCR diversity. J Immunol 174:7446–7452
Nossal GJ, Abbot A, Mitchell J, Lumus Z (1968) Antigens in immunity. XV. Ultrastructural features of antigen capture in primary and secondary lymphoid follicles. J Exp Med 127:277–290

Pawelec G, Barnett Y, Forsey R, Frasca D, Globerson A, McLeod J, Caruso C, Franceschi C, Fulop T, Gupta S, Mariani E, Mocchegiani E, Solana R (2002) T cells and aging. Front Biosci 7:1056–1183

Plowden J, Renshaw-Hoelscher M, Engleman C, Katz J, Sambhara S (2004) Innate immunity in aging: impact on macrophage function. Aging Cell 3:161–167

Qin D, Wu J, Caroll MJ, Burton GF, Szakal AK, Tew JG (1998) Evidence for an important interaction between a complement-derived CD21 ligand on follicular dendritic cells and CD21 on B cells in the initiation of IgG responses. J Immunol 161:4549–4554

Qin D, Wu J, Vora KA, Ravetch JV, Szakal AK, Manser T, Tew JG (2000) Fc gamma receptor IIB on follicular dendritic cells regulates the B cell recall response. J Immunol 164:6268–6275

Schnizlein CT, Szakal AK, Tew JG (1984) Follicular dendritic cells in the regulation and maintenance of immune responses. Immunobiology 168:391–402

Schwab R, Walters CA, Weksler ME (1989) Host defense mechanisms and aging. Semin Oncol 16:20–27

Song H, Price PW, Cerny J (1997) Age-related changes in antibody repertoire: contribution from T cells. Immunol Rev 160:55–62

Szakal AK, Hanna MG Jr (1968) The ultrastructure of antigen localization and viruslike particles in mouse spleen germinal centers. Exp Mol Pathol 8:75–89

Szakal AK, Taylor JK, Smith JP, Kosco MH, Burton GF, Tew JG (1990) Kinetics of germinal center development in lymph nodes of young and aging immune mice. Anat Rec 227:475–485

Szakal AK, Holmes KL, Tew JG (1983) Transport of immune complexes from the subcapsular sinus to lymph node follicles on the surface of nonphagocytic cells, including cells with dendritic morphology. J Immunol 131:1714–1727

Szakal AK, Kosco MH, Tew JG (1989) Microanatomy of lymphoid tissue during humoral immune responses: structure function relationships. Ann Rev Immunol 7:91–109

Szakal AK, Kosco MH, Tew JG (1988) A novel in vivo follicular dendritic cell-dependent iccosome-mediated mechanism for delivery of antigen to antigen-processing cells. J Immunol 140:341–353

Szakal AK, Aydar Y, Balogh P, Tew JG (2002) Molecular interactions of FDCs with B cells in aging. Semin Immunol 14:267–274

Szakal AK, Kapasi ZF, Masuda A, Tew JG (1992) Follicular dendritic cells in the alternative antigen transport pathway: microenvironment, cellular events, age and retrovirus related alterations. Semin Immunol 4:257–265

Tasat DR, Mancuso R, O'Connor S, Molinari B (2003) Age-dependent change in reactive oxygen species and nitric oxide generation by rat alveolar macrophages. Aging Cell 2:159–164

Taub DD, Longo DL (2005) Insights into thymic aging and regeneration. Immunol Rev 205:72–93

Taylor PR, Pickering MC, Kosco-Vilbois MH, Walport MJ, Botto M, Gordon S, Martinez-Pomares L (2002) The follicular dendritic cell restricted epitope, FDC-M2, is complement C4; localization of immune complexes in mouse tissues. Eur J Immunol 32:1888–1896

Tew JG, Wu J, Qin D, Helm S, Burton GF, Szakal AK (1997) Follicular dendritic cells and presentation of antigen and costimulatory signals to B cells. Immunol Rev 156:39–52

Tew JG, Wu J, Fakher M, Szakal AK, Qin D (2001) Follicular dendritic cells: beyond the necessity of T-cell help. Trends in Immunol 22:361–367

Tew JG, Mandel T, Burgess A, Hicks JD (1979) The antigen binding dendritic cell of the lymphoid follicles: evidence indicating its role in the maintenance and regulation of serum antibody levels. Adv Exp Med Biol 114:407–410

Tew JG, Kosco MH, Burton GF, Szakal AK (1990) Follicular dendritic cells as accessory cells. Immunol Rev 117:185–211

Tew JG, Kosco MH, Szakal AK (1989) The alternative antigen pathway. Immunol Today 10:229–232

Tew JG, Phipps RP, Mandel TE (1980) The maintenance and regulation of the humoral immune response: persisting antigen and the role of follicular antigen-binding dendritic cells as accessory cells. Immunol Rev 53:175–201

Thorbecke GJ, Lerman SP (1976) Germinal centers and their role in immune responses. In: Reichard SM, Escobar MR, Friedman H (eds) The reticuloendothelial system in health and disease: functions and characteristics. Plenum Press, New york, p 83
Thorbecke GL, Romano TJ, Lerman SP (1974) Regulatory mechanisms in proliferation and differentiation of lymphoid tissue, with particular reference to germinal center development. In: Brent L, Holbrow J (eds) Progress in immunology. North Holland Publishing Co, Amsterdam, pp 25–34
Villanueva JL, Solana R, Alonso MC, Pena J (1990) Changes in the expression of HLA-class II antigens on peripheral blood monocytes from aged humans. Dis Markers 8:85–91
Weksler ME, Szabo P (2000) The effect of age on the B-cell repertoire. J Clin Immunol 20:240–249
White RG, French VI, Stark JM (1970) A study of the localisation of a protein antigen in the chicken spleen and its relation to the formation of germinal centres. J Med Microbiol 3:65–83
Yang X, Stedra J, Cerny J (1996) Relative contribution of T and B cells to hypermutation and selection of the antibody repertoire in germinal centers of aged mice. J Exp Med 183:959–970

# Index




P. Balogh (ed.), *Developmental Biology of Peripheral Lymphoid Organs*,
DOI 10.1007/978-3-642-14429-5, © Springer-Verlag Berlin Heidelberg 2011

DEPAUL UNIVERSITY
3 0511 01037 4098